ENERGY-EFFICIENT HOUSING DESIGN

ENERGY-EFFICIENT HOUSING DESIGN

A Combined Approach

Jonathan Lane

VNR VAN NOSTRAND REINHOLD COMPANY
New York

Renderings: figs. 4-2, 4-6, 4-8, 4-12, 4-13, 4-15, 17-7, 17-14—Mark W. Baechtle

Library of Congress Catalog Card Number 85-17906
ISBN 0-442-26033-4

Printed in the United States of America
Designed by Loudan Enterprises

Van Nostrand Reinhold Company Inc.
115 Fifth Avenue
New York, New York 10003

Van Nostrand Reinhold Company Limited
Molly Millars Lane
Wokingham, Berkshire RG11 2PY, England

Van Nostrand Reinhold
480 La Trobe Street
Melbourne, Victoria 3000, Australia

Macmillan of Canada
Division of Canada Publishing Corporation
164 Commander Boulevard
Agincourt, Ontario M1S 3C7, Canada

16 15 14 13 12 11 10 9 8 7 6 5 4 3 2 1

Library of Congress Cataloging-in-Publication Data
Lane, Jonathan, 1931–
Energy-efficient housing design.
Bibliography: p.
1. Architecture, Domestic—United States.
2. Architecture and energy conservation—United States.
3. Solar houses—United States. I. Title.
NA7117.3.L36 1986 728.3'7'0470973
85-17906
ISBN 0-442-26033-4

Contents

Preface

The single-family house represents the realization of the American dream. Tens of millions of single-family houses have been built in this country since the end of World War II. In the past few years, however, the cost of owning a home has soared dramatically. The cost of purchasing a home has increased much more rapidly than inflation, because of rising land costs, more stringent regulation of development, and large increases in the price of construction materials. Mortgage rates have risen even more drastically, going from 6 percent in the late 1960s to 12 to 14 percent in the early 1980s. The final element in the unfavorable cost equation has been increased energy prices. The single-family house uses large amounts of energy, and spiraling charges for gas, oil, and electricity have added to the burden imposed by higher prices and higher mortgage rates.

Mortgage rates and, to a great extent, house prices, cannot be brought down by the architect and engineer. Energy consumption can, however, be significantly reduced by design changes. Since the first sharp rise in energy prices in the early 1970s, a major effort has therefore been made to construct buildings that consume less energy. Initially, major attention was centered on active solar systems. More recently, the search for the energy-efficient home has focused on passive solar systems, earth sheltered designs, and the use of very heavy insulation. Many of the major technical problems of each of these design concepts have been solved, and new refinements continue to be introduced. The obvious advantages of combining the three methods of reducing energy consumption have, however, been largely overlooked.

The combination of earth shelter, passive solar, and superinsulation is a very natural approach to cost-effective, energy-efficient housing design. Combining the three approaches allows each to reinforce the others to create a more effective overall design. A partially earth sheltered building can be heavily insulated more easily than the conventional frame house, and at the same time, combining partial earth cover with superinsulation shields the house from the wind and provides extra thermal protection. With heating loads greatly reduced by superinsulation and by the use of earth shelter, windows of little more than ordinary size can, if they are oriented toward the sun, provide almost all the energy needed for winter space heating. In warm weather, superinsulation and earth cover reduce cooling requirements; throughout the year the passive solar system can also provide energy for domestic hot-water heating. The combined approach greatly reduces energy costs with only a modest increase in the cost of constructing the building. The result is a house that pays for itself from the start and if even a very moderate rate of inflation is projected for energy prices, very large savings occur in future years.

The study of the earth sheltered/passive solar/superinsulated (ESSPS) house has led, rather unexpectedly, beyond the energy question and problems of building design to the subject of suburban subdivision planning. If energy-efficient houses are to be built in large numbers, they must be constructed as mass-built tract homes. The unconventional if simply built houses proposed here are very different from the standard tract house in their relationship to the building lot and to the street. The character of the tract plan changes with the energy-efficient house, and new types of tract layouts must be developed in order to meet its requirements.

In the energy-efficient subdivision, the partially earth sheltered house, set behind artificial berms or fitted into the slope of the land, blends into its site. In place of long rows of one- or two-story houses, standing side by side along the street, there will be protected courtyards, with houses virtually concealed from view or only partially visible, and the street vista will now be dominated not by buildings but by the gentler forms of tree lines, plant masses, and the shaped or naturally contoured land.

Note on Illustrative Examples

The buildings illustrated in this book are designed to be built in southeastern Pennsylvania, a region with 5,100 heating degree days and 500 cooling degree days. In the northern half of the United States, heating degree days vary from 3,000 to 8,000, and the variation in annual cooling demand is as great. Local areas also show great differences in the availability of winter sunlight, which affects the design of passive solar systems. Still other local factors, including construction costs and current and estimated future prices of electricity, gas, and fuel oil must be taken into account. The amount of insulation used in the cost-effective, energy-efficient house, as well as the window size and the capacity of the passive solar storage system, will therefore vary regionally and locally.

PART I

BUILDING DESIGN

1

Current Approaches to Energy-efficient Design

Active and Passive Solar Systems

The search for the cost-effective, energy-efficient house has been characterized by solutions that approach the problem of energy efficiency in fundamentally different ways. Active solar, the first approach to be intensively developed in the United States, sought a high tech solution. Passive solar and earth sheltered design, two approaches that have proven more successful than active solar, rely primarily on redesigning the house, rather than on major technological innovations. Superinsulation, the newest approach, is the product of modification of a standard house type, the familiar frame dwelling, rather than of the dramatic new design concepts that characterize the passive solar and earth sheltered solutions.

The basic active solar system, with its rooftop collectors and basement storage facility, was first developed more than twenty-five years before the effects of the energy crisis of the 1970s were felt. As home heating costs rose swiftly in the middle and late 1970s, a large-scale effort was made to turn this experimental device into a practical heating system. Over a period of almost ten years, all aspects of the design of the active system were thoroughly explored.

When the research effort ended in the early 1980s, the active system retained its original form. No major breakthroughs had been made, and, although the basic design had been greatly refined, no fundamental design changes were introduced. The refined system was clearly workable, but it remained economically unusable despite the great increase in energy prices during the preceding decade. The cost of a system capable of heating a house of average size had not been brought below the $35,000 to $40,000 range, and installing an active system substantially increased the homeowner's monthly mortgage bill—an increase that was several times larger than the savings in fuel costs.

Advocates of active solar, anticipating continued rapid escalation of energy costs, maintained that the expensive system was cost-effective, because it would eventually pay for itself as energy prices continued to rise. For the typical home buyer, however, the active solar installation created a very large, immediate financial burden, whereas savings in energy costs would only be obtained many years in the future. Experience with active solar demonstrated that technically feasible solutions would be practical only if they were also cost-effective.

Once it became clear that research breakthroughs would not lead to radical reductions in the cost of active solar, the passive

A. **Active Solar**

B. **Passive Solar**

C. **Earth Sheltered House**

solar concept attracted widespread attention. Passive solar design uses the living spaces of the house to gather energy from the sun, rather than an elaborate set of rooftop collectors, and building mass provides a large part of the system's storage capacity. In addition to being much simpler technically than active solar, passive solar encouraged innovative architecture. Use of large, south-facing glass areas to collect solar energy led naturally to designs with open, sunfilled living spaces. Many homes also included a greenhouse, used not only for growing ornamental plants but for producing food.

D. **Superinsulated House**

Passive solar, though much simpler and more practical than active solar, also has significant design limitations. The simple passive collector, made up of large walls of insulating glass, is a very inefficient device, because a high proportion of the energy acquired on bright days is lost by radiation at night and during bad weather; only a small percentage is retained as useful gain. A second problem is that storing energy that can later be released to the building requires large internal temperature swings. Even with temperature swings of fifteen to twenty degrees Fahrenheit, each unit of storage mass can acquire only a relatively small amount of energy, so that in addition to its building mass the passive solar house requires an elaborate storage system, such as a deep rock bin under the building's slab. Summer

1–1. Four approaches to energy-efficient design. (C reprinted from Underground Space Center, *Earth Sheltered Homes,* p. 118; D reprinted, by permission, from Argue, *The Well-Tempered House,* p. 163.)

cooling is also a problem. Passive solar relies on large south-facing glass areas, and the sun load on the glass makes it difficult to maintain comfortable temperatures in summer.

Many different solutions to the problems inherent in passive solar have been proposed. One of the most cost-effective refinements is the provision of heavy thermal protection for the building, using extra insulation and, in some cases, limited earth shelter. When thermal protection is increased, the load on the passive system drops, and the size and cost of the collector and storage components can be reduced. Extra thermal protection is, however, usually regarded as an auxiliary design characteristic of the passive solar house.

Earth Sheltered and Superinsulated Houses

Earth sheltered architecture is the most radical of the currently proposed departures from conventional building design. The earth sheltered structure blends into the landscape, and from most angles becomes invisible, appearing as either a low, covered mound, or when built on a sloping site, as an extension of the hillside. The structure has minimal impact on the site and conveys a sense of well-protected interior space. The earth sheltered concept has produced some remarkably handsome designs, while effectively reducing energy requirements for both heating and cooling by combining very substantial thermal protection with the airtight envelope created by the earth cover on the building. Many earth sheltered houses also incorporate a passive solar element. Passive solar can be added very easily, because most earth sheltered houses have concrete floor slabs and masonry walls that provide a large amount of built-in storage capacity.

Deep earth cover on the roof of the houses requires a massive support structure; most earth sheltered houses are built with minimal earth cover. The cover is typically only 12 to 18 inches deep and is backed up by 2 to 6 inches of rigid insulation; rigid insulation is also applied to the walls. But even with minimal earth cover, total roof loads reach 200 to 300 pounds per square foot, compared to 35 to 45 pounds per square foot with conventional construction. This requires a heavy and expensive roof structure that must be supported at frequent intervals by block cross walls or sturdy posts. The earth-covered roof must also be carefully waterproofed, since roof leaks under its earth cover are extremely difficult to locate and repair. Installation of waterproofing durable enough to eliminate the danger of leaks can be a major expense, and, together with the cost of the roof structure and roof supports, adds significantly to the overall cost of earth sheltered construction.

Building with minimal earth cover, of not more than 12 to 18 inches, limits thermal protection for the house; a thin layer of earth has comparatively little insulation value. The rigid insulation used to supplement the earth cover provides most of the roof's protection against heat loss, and the whole roof typically has a transmission rating of R-12 to R-30. This is comparable to 3 to 9 inches of batt insulation, and the principal advantage of the earth sheltered roof is its ability to provide an airtight cover for the building. Similar problems are not encountered in wall construction, since the earth sheltered wall is comparatively easy to construct and its deeper earth cover gives more substantial thermal protection to the building.

Superinsulation is another method of reducing the space-heating demand of a building. The superinsulated house is so prosaic that it has received much less attention than the other approaches to energy efficiency. Frame construction is modified to incorporate extremely heavy thermal protection and an airtight membrane. The airtight membrane is needed if the thermal protection is to be used properly, since 35 to 40 percent of the heat loss in a well-built frame house results from air infiltration. Unless air infiltration is controlled, using large amounts of insulation is not cost-effective; the extra insulation produces relatively small energy savings. If the building is sealed, walls with ratings as high as R-50 become practical, while roof ratings can reach R-60 or, under some conditions, even higher.

The sealed superinsulated house requires mechanical ventilation in order to provide an adequate amount of fresh air for its occupants; without proper ventilation the sealed house can quickly become uninhabitable. Superinsulated houses are equipped with heat exchangers that use the heat of the stale air exhausted by the ventilation system to warm cold outside air as it enters the building. The heat exchanger both ensures the habitability of the superinsulated house and improves its cost-effectiveness.

The combination of a heavy thermal envelope, an infiltration barrier, and the heat exchanger creates savings of 50 to 70 percent in the annual demand for space heating, while providing a minor reduction in summer cooling loads. Other approaches to energy efficiency can often produce greater overall energy savings, but superinsulation is the most cost-effective approach developed to date.

2

Superinsulation Methods

Insulation and Superinsulation

During the past ten years, insulation standards for new homes have been substantially upgraded. Roof insulation has gone from R-11 (3½ inches) to the R-19 to R-30 range (6 to 9 inches). Wall insulation values

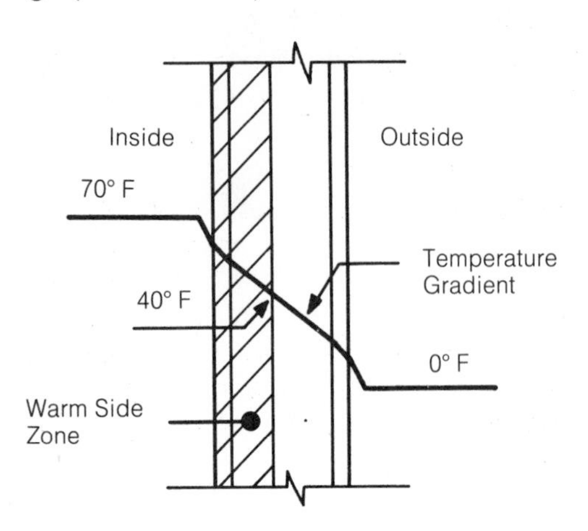

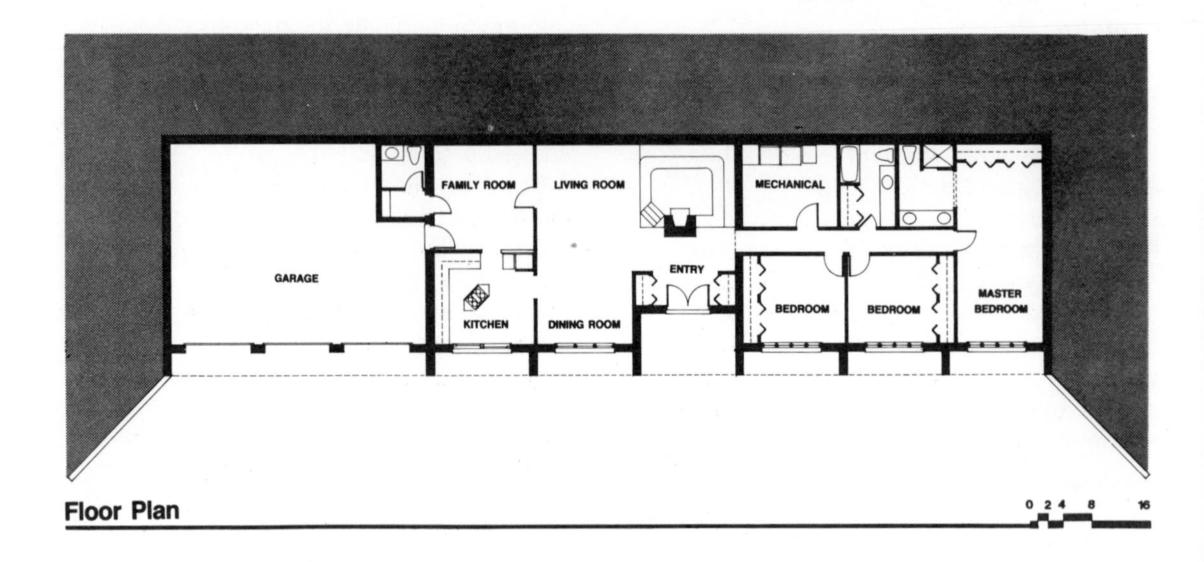

2–1. Exterior wall temperature gradients. (Floor plan reprinted from Underground Space Center, *Earth Sheltered Homes,* p. 119.)

have not been increased as much, because the insulation thickness is restricted by stud depth. In many areas, however, the R-11 batts that 2 × 4 studs can accept have been supplemented by 1-inch polystyrene sheathing, which has a thermal resistance of R-5.

Superinsulation goes beyond these standards of thermal protection and makes efficient use of its heavier thermal envelope by combining it with an airtight membrane that eliminates the transmission of water vapor from the interior of the house, while preventing cold outside air from entering the building. Without this membrane, water vapor escapes outward from the heated interior in winter and condenses and freezes within the walls as it reaches their cold side (figure 2–1). This process degrades the performance of any insulated building over time. As the thickness of the insulation is increased the problem becomes more serious, and control of vapor transmission as well as air infiltration becomes necessary. In order to perform properly, the transmission and air infiltration barrier must provide a tight envelope around the living spaces of the house. A tight seal around the house is particularly important for control of vapor transmission.

Superinsulated walls are built in varying thicknesses, depending on climatic conditions. *Single-layer* walls combine the membrane with insulation in the range of R-19 to R-24; *multilayer* walls provide heavier thermal protection, and ratings with the multilayer wall can reach R-50.

Superinsulated Construction

Single-layer frame superinsulated construction is suitable for regions with winter heating seasons of up to 7,000 degree days. The single-shell wall is usually built with 2 × 6

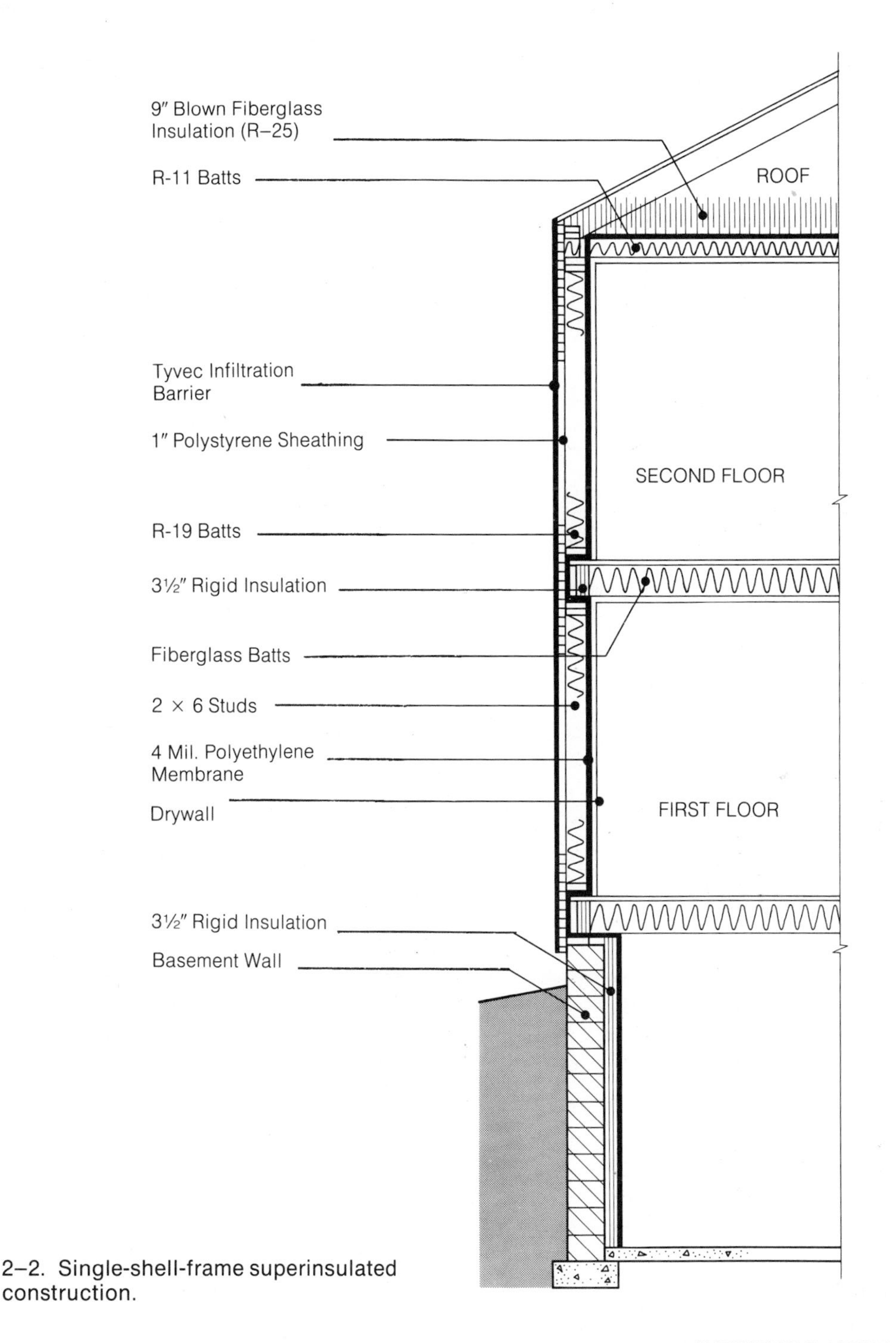

2–2. Single-shell-frame superinsulated construction.

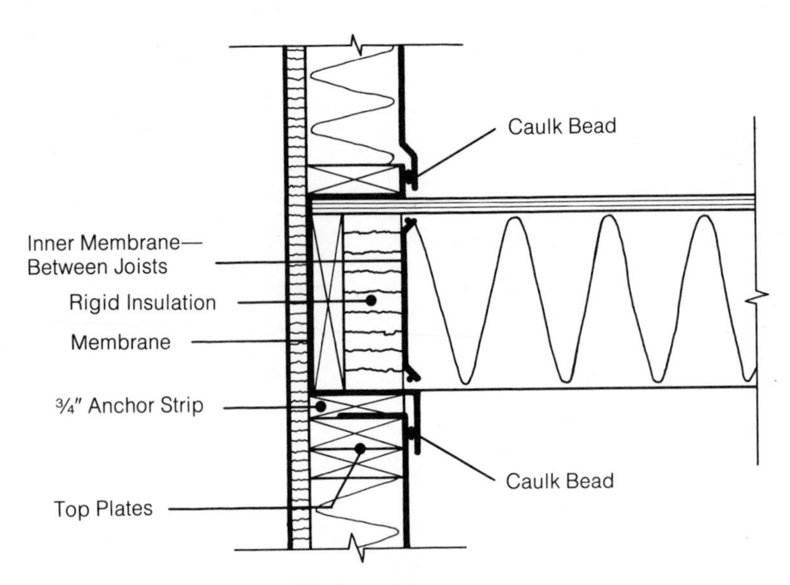

DETAIL AT SECOND-FLOOR JOISTS

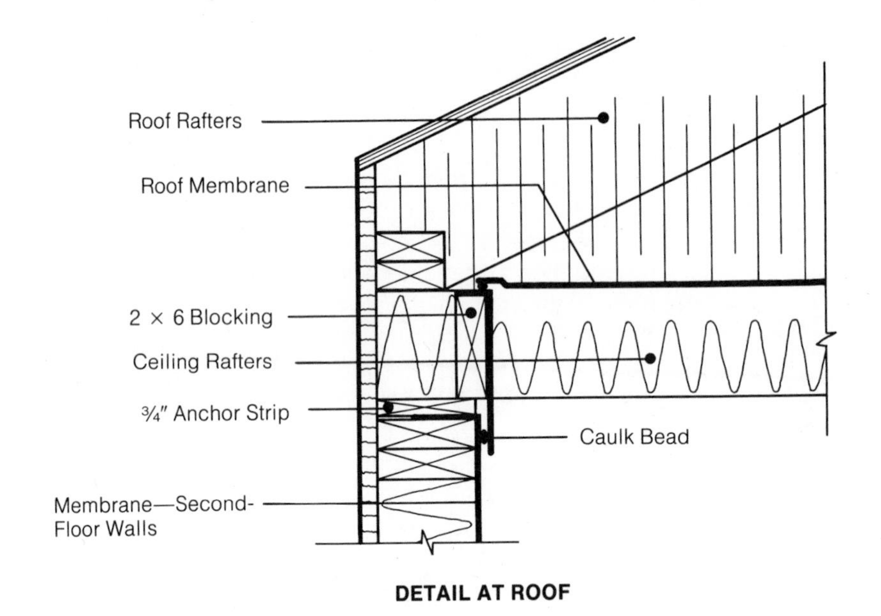

DETAIL AT ROOF

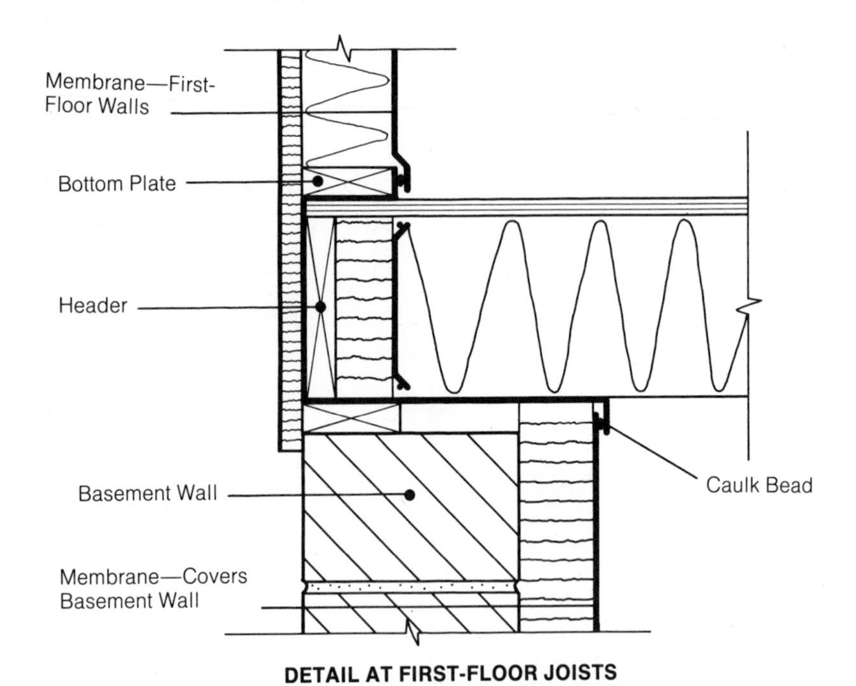

DETAIL AT FIRST-FLOOR JOISTS

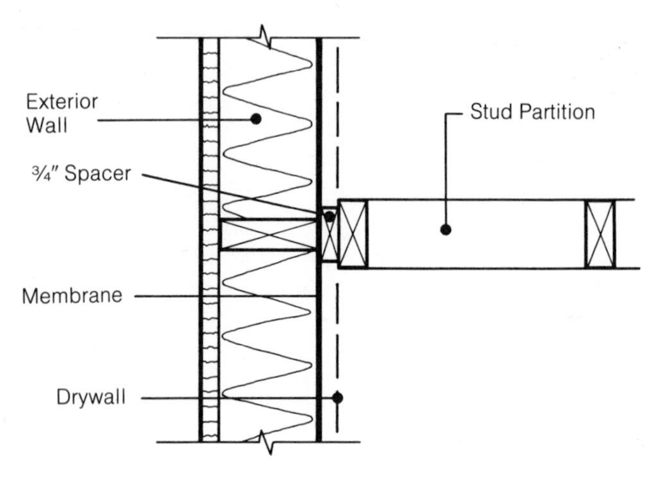

JOINT AT INTERIOR PARTITION

2–3. Details of single-shell superinsulated construction.

studs to accommodate R-19 batts; the addition of polystyrene sheathing brings total thermal protection to R-24. The infiltration barrier is created by placing 4-millimeter polyethylene sheets on the inside face of the studs, so that they are on the warm side of the wall.

Construction of the walls is complicated significantly by the addition of the membrane. Single-layer superinsulated construction requires careful detailing and use of appropriate construction procedures. The membrane of the single-layer superinsulated house is frequently stapled to the studs after the outside walls are up. Polyethylene forms a very effective vapor and infiltration barrier, but it is extremely difficult to handle in large sheets; it is therefore easier and more efficient to apply the material to the walls when they are still on the deck. In order to apply polyethylene this way, the assembled walls must be turned over once the studs have been nailed together and exterior sheathing has been applied. The membrane material is then rolled out lengthwise over the studs and firmly secured by a 1 × 4-inch strip nailed to the top plate (figure 2–3). The sheets should not be unfolded at this point, but rather tacked in place still folded in their original 18-inch or 24-inch width. This allows the walls to be braced as they are set up and permits the construction of bearing partitions and placement of insulation between the studs.

Bearing partitions are installed once the outside walls are up and braced. In order to avoid a gap in the membrane, which is still folded, full-height strips of polyethylene are installed at the ends of the bearing partitions before they are erected. The membrane is unrolled from the top of the walls, after the load-bearing walls are up and insulation has been installed, and cut out around the bearing partitions. The separate pieces are then sealed to the polyethylene strips placed at the partitions earlier. This can be done either with a caulk joint, or by using a heat-sealing tool. Lap joints at the corners of the house are also needed. Nonbearing partitions should be erected after the membrane is unrolled in order to reduce the number of joints that must be formed. Nonbearing partitions require spacers at their ends to allow enough room to set up the partitions without damaging the membrane.

Detailing the membrane at the floor joists is especially important. The joists must be wrapped with separate strips of polyethylene, which are placed over the foundation top plate or the top plate of the first floor partitions before the joists are laid down. These strips are needed in order to avoid a break in the windproofing, but are located on the outside of the wall and cannot act as a vapor barrier. At least 2 inches of rigid insulation must be cut in between the joists, backed by a second, warm-side membrane layer. This layer is made up of small pieces of polyethylene pressed into place against caulk beads that run around the four sides of the joist bays.

All of the multiple joints of the membrane must be carefully formed, since a typical two-story house will have over 800 linear feet of exposed joints (figure 2–4). Penetration of the membrane with plumbing, electrical, and heating systems must also be avoided wherever possible. Electrical outlets on the exte-

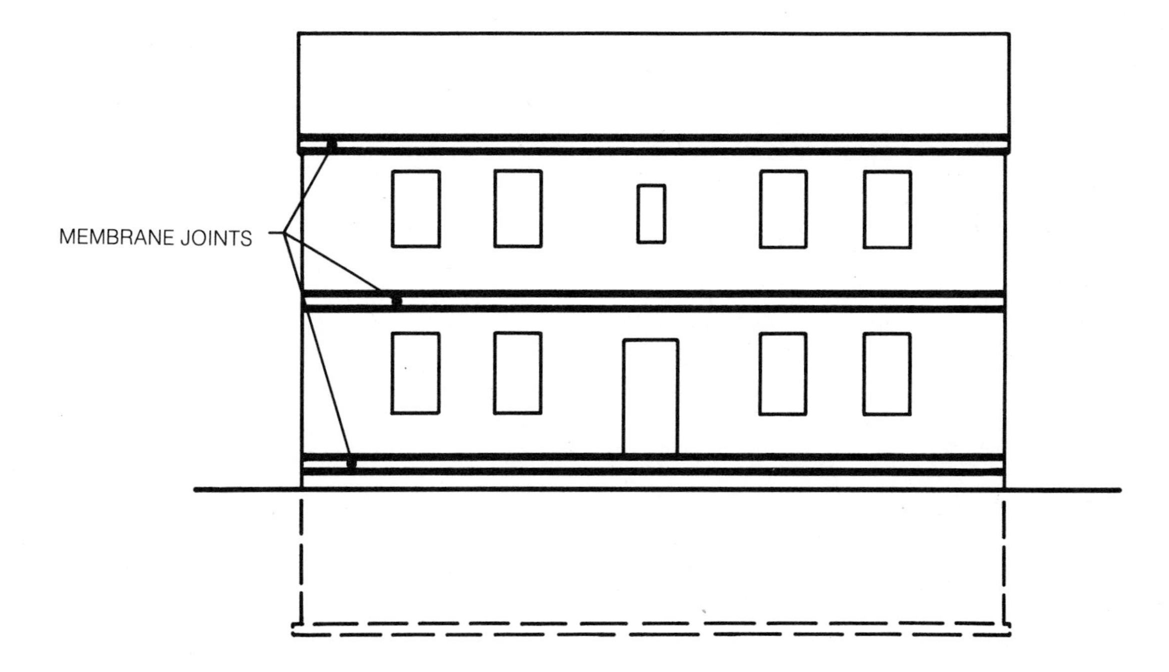

2–4. Membrane joints in a two-story frame house.

rior walls will create a large number of openings in the membrane; to keep these openings sealed, the outlets should be housed in metal pans, which are caulked around their edges. Waterproof floor outlets can be substituted for the wall fittings.

The multilayer superinsulated wall is much more elaborate than the single-layer wall, but will produce much larger energy savings in severe climates. Development of the multilayer wall has been influenced by the Saskatchewan Conservation House, constructed in Canada in 1980. Intended for a very severe climate (10,700 heating degree days), this building has double-stud walls that are separated by a third layer of insulation (figure 2–5). The membrane is located on the outside face of the inner shell, which allows electrical wiring to be placed in the walls without penetrating the membrane and eliminates the risk of damaging the membrane during construction. Two-thirds of the insulation is located on the cold side of the vapor barrier, and escaping water vapor cannot condense or freeze before it reaches the membrane.

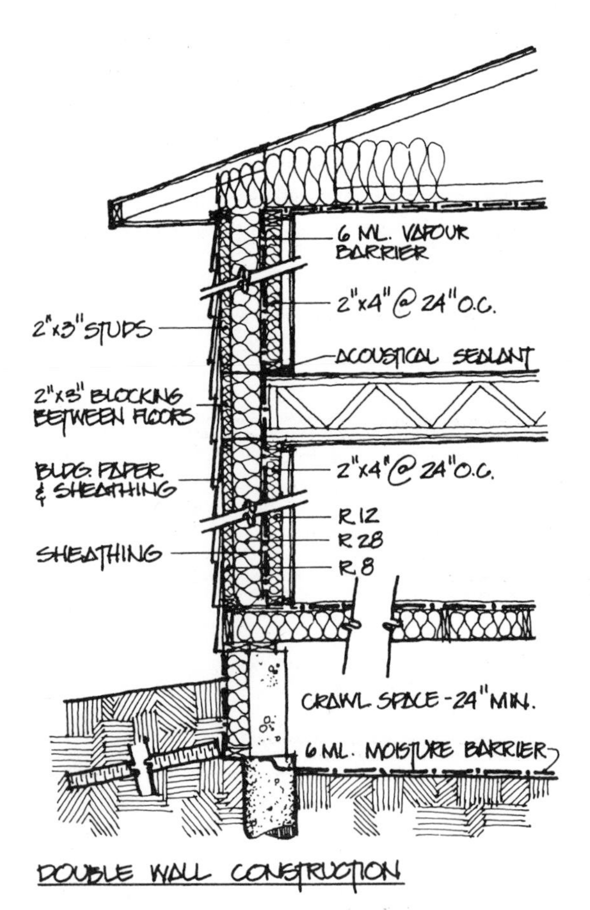

2–5. Wall section of Saskatchewan Conservation House. (Reprinted, by permission, from Argue, *The Well-Tempered House,* p. 143.)

The inner and outer stud walls of the house were laid out together on the floor decks to simplify construction. After assembly of the studs, plywood plates were nailed to the top and bottom of the wall sections, separating them and allowing the entire wall to be erected as a single unit. Constructing the multilayer wall requires, however, a large number of separate steps. Once the two stud shells are assembled, they must be moved apart, sheathed, and insulated. The polyethylene liner must also be installed before final assembly. In addition, separate knee walls are required at each floor to make up the thickness of the wall, and the main walls must be built in relatively short sections because of their weight. The complexity of multilayer-frame superinsulated construction has led to the introduction of factory-built wall units, which eliminate the cumbersome field assembly process.

Roof construction of the superinsulated house is much simpler, because the conventional A-roof can, with little modification, become a superinsulated roof structure capable of providing almost any amount of thermal protection. Membrane location is important. It is difficult to install the membrane by nailing it to the bottom of the ceiling rafters, and a hung membrane will have joints at the bearing partitions and will be penetrated by wiring runs and electrical outlets. These problems can, however, be avoided if the membrane is placed on top of the ceiling rafters. It is installed very easily in this location by rolling the polyethylene sheets out across the rafters and then unfolding them. Insulation is applied in two layers. Batt insulation is placed between the joists. Additional batts or blown insulation forms the second layer, which goes on top of the membrane. In order to keep the membrane in the warm zone, the upper insulation layer must have a minimum resistance of R-22 when R-11 batts are used between the rafters, and R-38 resistance when R-19 batts are used for the inner layer of insulation. This results in minimum roof ratings of R-33 and R-57, but extra insulation can be added to the top of the membrane, bringing the roof rating to any desired level.

Superinsulated houses are also built with panels of expanded polystyrene insulation (EPS), replacing the familiar stud-and-fiberglass-batt construction. The EPS shell normally does not require incorporation of a vapor barrier, because it is much less sensitive to moisture than fiberglass and other low-density insulation materials. EPS insulation has very low permeability to moisture; when small quantities of moisture do reach the cold side of the wall, they have only a minor effect on the insulating properties of the material. In very cold climates, however, it is often desirable to apply vapor resistant paint to the inside of the panels to form a partial vapor barrier.

EPS panels can be joined with a vapor-proof foam adhesive, so that a tight shell is formed that is highly resistant to both wind and moisture. The panels usually have plywood sheathing on the outside of the insulation to form a nailing base for the exterior finish, and sheetrock is laminated to the inside face of the insulation. Nonstructural panels are often used with post-and-beam construction, and panels with integral structural members have also been developed. Insulation thickness for wall and roof panels varies from 3½ inches to 7½ inches, with insulation values of R-17 to R-37. Insulation ratings are usually lower than those of the conventionally framed superinsulated house, but the EPS system makes it much easier to achieve airtight construction and emphasis is shifted still more strongly toward achieving energy savings by controlling air infiltration.

The panelized EPS structure avoids the problem of membrane placement, which complicates the construction of the conventionally framed superinsulated house. The trade-off is use of more expensive insulation materials, which limit practical R-values, as well as more expensive framing methods: the economical job-built stud wall is replaced either by costly post-and-beam construction, or by elaborate prefabricated structural panels.

3

Combining Approaches: The Earth Sheltered/Superinsulated/Passive Solar House

Superinsulation requires complicated construction methods to create a thick, windproof thermal envelope for the frame house. A more appropriate superinsulated building structure can be developed that is free of excessive complication and allows the integration of other approaches to energy-efficient housing, including earth shelter and passive solar. The three approaches complement each other and, when used together, result in a simpler, more cost-effective, and more energy-efficient design than is possible with superinsulation alone.

Residential energy consumption has four components: winter space heating, summer cooling, hot-water heating, and electricity, used for lights and appliances. The ESSPS house is, like most energy-efficient buildings, designed to reduce sharply the cost of winter space heating. It also reduces the amount of energy used for summer cooling and uses surplus passive solar energy for hot-water preheat. Space heating costs can be reduced by as much as 90 percent, cooling costs by nearly 50 percent, and there can be a 40-percent savings in the cost of hot-water heating.

The energy demand of space heating the ESSPS house is reduced by its combination of superinsulation and partial earth shelter, which together provide very heavy thermal protection and effective resistance to air infiltration. The passive solar system provides almost all of the remaining requirement for space-heating energy. Heavy thermal protection also serves to reduce summer cooling loads, as do the sun-control devices that are incorporated in the design. Surplus passive energy is available throughout the year and is used to preheat hot water, after which a conventional (or solar) heater can bring it to its final temperature. In summer, the preheat system becomes a heat sink, which helps cool the house, as cold water brought into the domestic hot-water system absorbs heat from the building space.

Space heating, cooling, and hot-water heating costs for the ESSPS house will typically be reduced by 80 percent in comparison with a well-built conventional house insulated to currently accepted industry standards. Savings do not extend to the fourth component of energy consumption, electricity used for purposes other than cooling and hot-water heating. Savings in this area cannot, at present, be achieved by modifying the building design, except

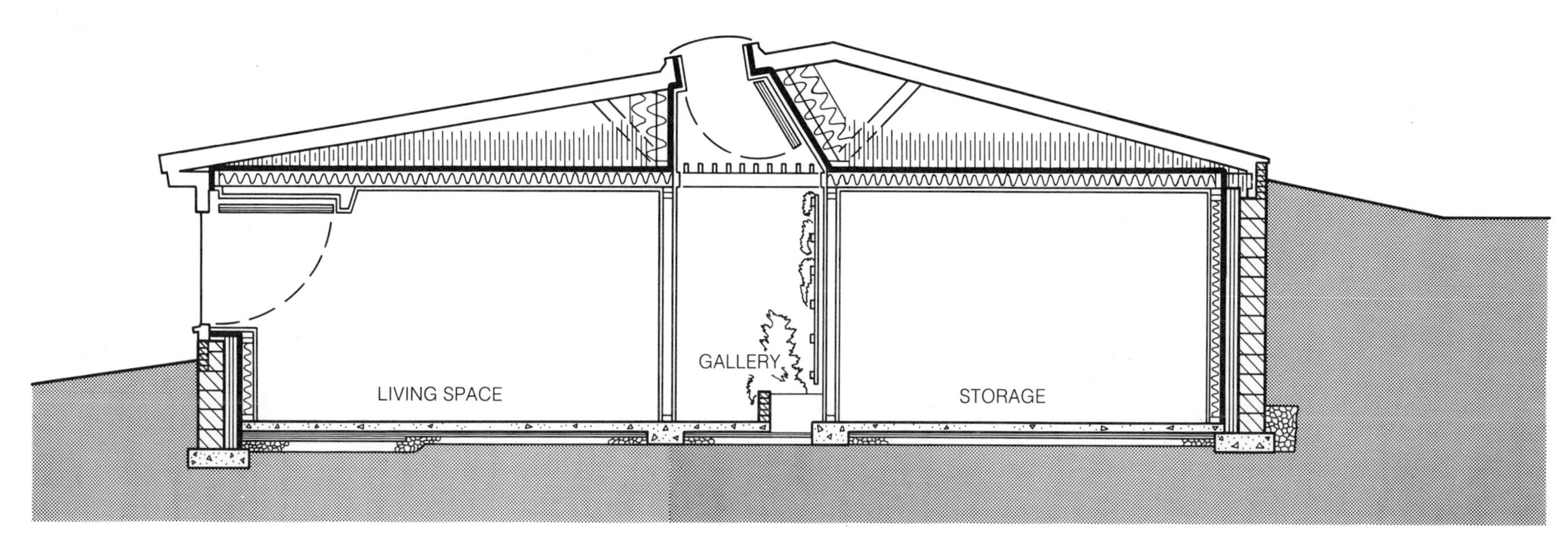

3–1. Cross section of the ESSPS house.

through installing energy-efficient appliances and lighting equipment.

The Building Section

A cross section of the ESSPS house shows a one-story building with earth sheltered masonry walls, a heavily insulated A-roof, and a concrete floor slab (figure 3–1). Most of the living spaces are placed on the south side of the house to obtain adequate passive solar gain, and the closed north side is used primarily for service spaces, including entrance and garage as well as storage and utility areas. The building has wide, skylit corridors, which separate the living and service areas.

Roof construction is generally similar to that of the superinsulated A-roof described in Chapter 2, *Superinsulated Construction.* This roof is placed over walls designed to combine masonry construction and earth shelter with multiple layers of insulation and a polyethylene infiltration barrier. The heavy insulation is continued under the perimeter of the slab. Thermal shutters over the windows complete the insulation envelope of the house and allow efficient acquisition of passive gain.

Walls

The earth sheltered walls form a multilayer shell that gives thermal protection equal to or greater than frame superinsulation with much simpler construction. The block outer shell of the ESSPS wall is its structural component, providing both support for the roof and resistance to the pressure of the surrounding earth. The block is lined with a thick layer of rigid insulation, which in turn is covered by the polyethylene infiltration barrier. An inner shell is placed on the warm side of the membrane; its studs carry the interior finish and provide space for additional insulation.

Heavy insulation inside the block wall is required because the 8 feet of earth cover protecting the building will only provide an average thermal protection of R-9; this protection will vary from approximately R-2—close to the surface—to R-16—at the bottom of the wall. But in addition to providing thermal resistance for the walls that averages just slightly below the resistance provided by the standard 3½-inch fiberglass batt, the earth cover protects the house from air infiltration, gives very substantial thermal protection to the edges of the concrete slab, and lowers construction costs by greatly reducing the amount of foundation work and the size of the exposed wall area.

The wall design keeps the infiltration barrier well inside the warm zone in a location that is free of penetration by electrical outlets, wiring, and nails and eliminates the risk

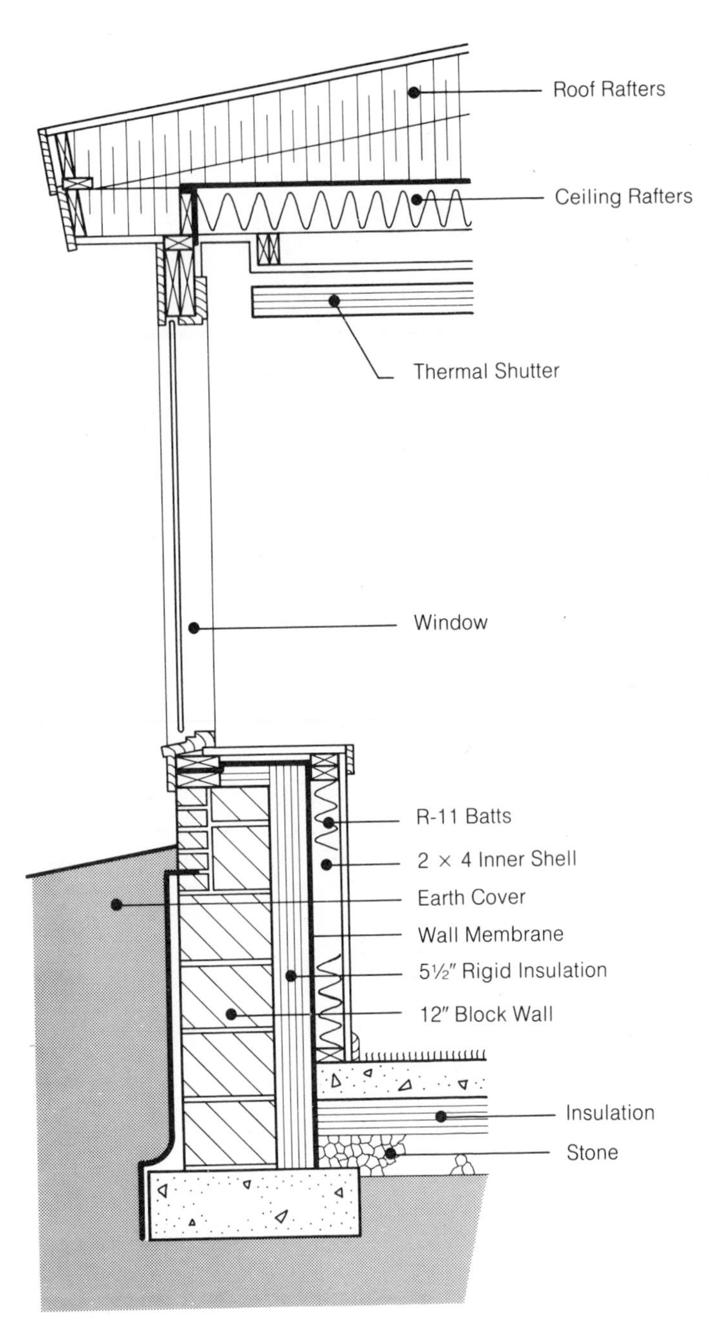

WINDOW WALL

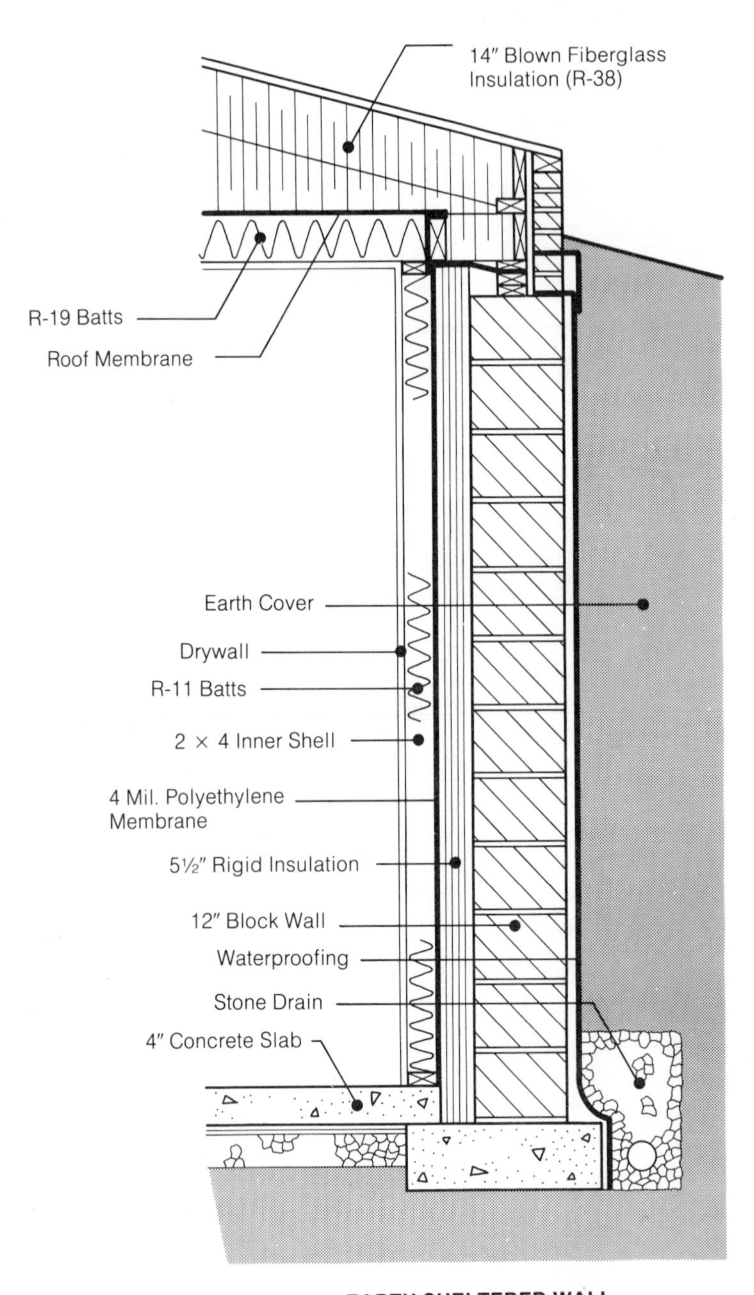

EARTH SHELTERED WALL

3–2. Wall sections of the ESSPS house.

of construction damage. There are no restrictions on the amount of insulation that can be installed, and insulation thickness will depend on cost-effectiveness in different climates. Table 3–1 shows suggested wall insulation thicknesses.

Its multilayer construction makes the heavily insulated, airtight walls of the ESSPS house very easy to build (figure 3–3). Each layer of the wall can be completed separately, so that each trade can complete its work in one operation. The concrete or block walls that form the outer shell are constructed first. Rigid insulation is then laminated to the structural shell to form the middle layer of the wall, and the polyethylene membrane, secured to the top plate on the block wall, is rolled out over the inner face of the insulation and tacked in place at the bottom of the wall. The final stage is framing and insulating the inner shell. In addition to providing heavy thermal protection, the multilayer wall is airtight, because its membrane extends, without a break, from the top of the wall to the bottom, where the joint with the slab is completely protected from the wind. Vertical joints occur only at the corners of the house and are protected over most of their length from air infiltration by the earth cover. The junction between the walls and the roof creates the only exposed horizontal joint. Reducing the number of membrane seams greatly simplifies the construction process.

Table 3–1. Insulation Thickness and Thermal Resistance Ratings of ESSPS Walls

Climate	Batt Insulation		Rigid Insulation		Earth Shelter	Total Thermal Resistance
	Thickness	R Value	Thickness	R Value	Av. R Value	
Moderate	2½″	R-8	3½″	R-17	R-9	R-34
Severe	3½″	R-11	5½″	R-27	R-9	R-47
Very severe	6″	R-19	7½″	R-37	R-9	R-63

Table 3–2. Insulation Thickness and Thermal Resistance Ratings of ESSPS Walls with Laminated Expanded Polystyrene Panels

Climate	EPS Insulation		Earth Shelter	Total Thermal Resistance
	Thickness	R Value	Av. R Value	
Moderate	5½″	R-27	9	R-38
Severe/very severe	7½″	R-37	9	R-47

Laminated panels of expanded polystyrene can also be used for the walls of the ESSPS house. Drywall is laminated to the inside face of the insulation board placed inside the structural block walls, and the membrane and inner stud shell are eliminated. In moderate climates, 5½-inch-thick EPS panels are used; in severe and very severe climates 7½-inch-thick EPS panels are required.

The panels are secured at the top and bottom with 2 × 6 or 2 × 8 plates; joints between the panels are sealed with vaporproof adhesive foam. Vapor resistant paint should be applied to the inner face of the panels, even in moderate climates, because construction is partly below grade.

The laminated panels simplify basic construction, although installation of electrical wiring is difficult and requires channeling the panels for electrical runs. The overall cost of the two- and three-layer walls will be nearly the same. The three-layer wall requires more labor but uses less material. The laminated EPS panel will often prove somewhat more economical in areas where carpenters are scarce and expensive; where an efficient framing crew is available and labor rates are moderate, the three-layer wall will cost less to build.

Insulating blocks, made with expanded polystyrene, have also been developed and can be used to form a one-layer wall system for the ESSPS house. Manufactured in Canada, as yet only at a single plant, the "Sparfil" blocks are formed of granules of expanded polystyrene bonded together in a cement mix. The hollow blocks are filled with foam EPS inserts. Twelve-inch blocks, with inserts, have a thermal resistance of R-24.5, nearly as high as the rating achieved by lining conventional block with

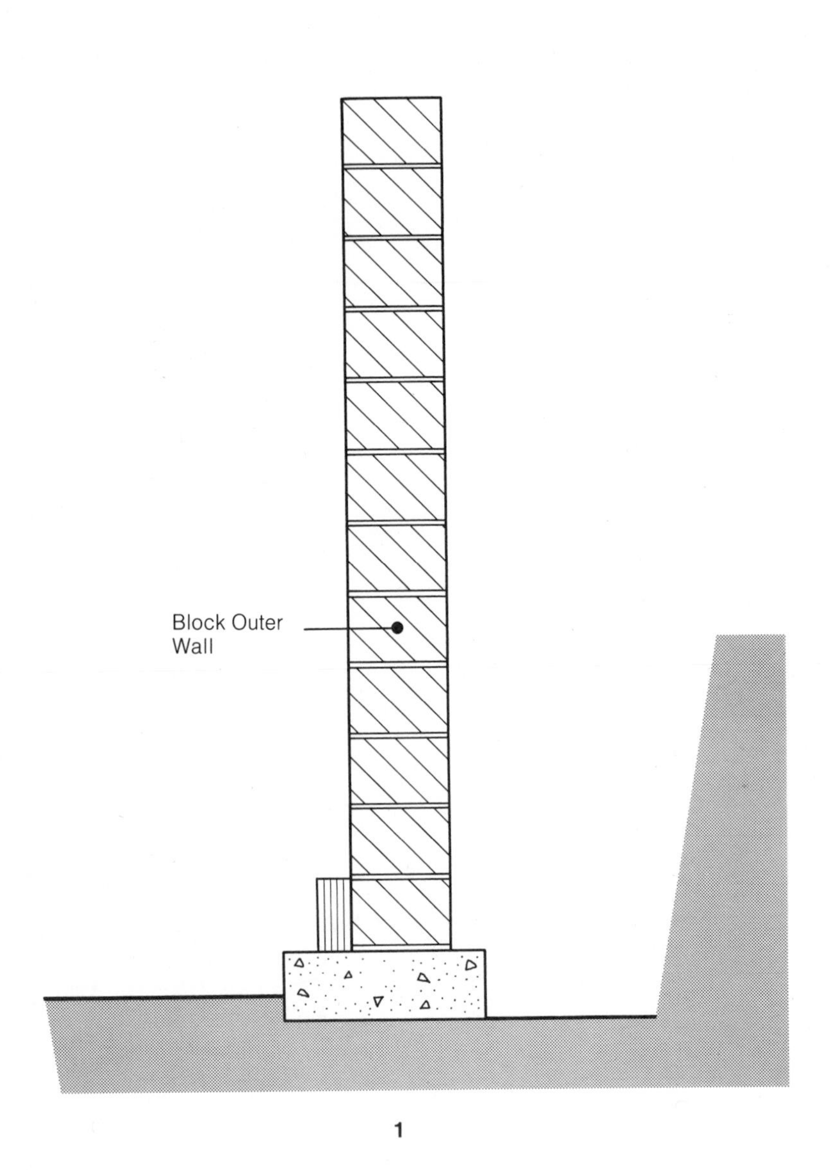

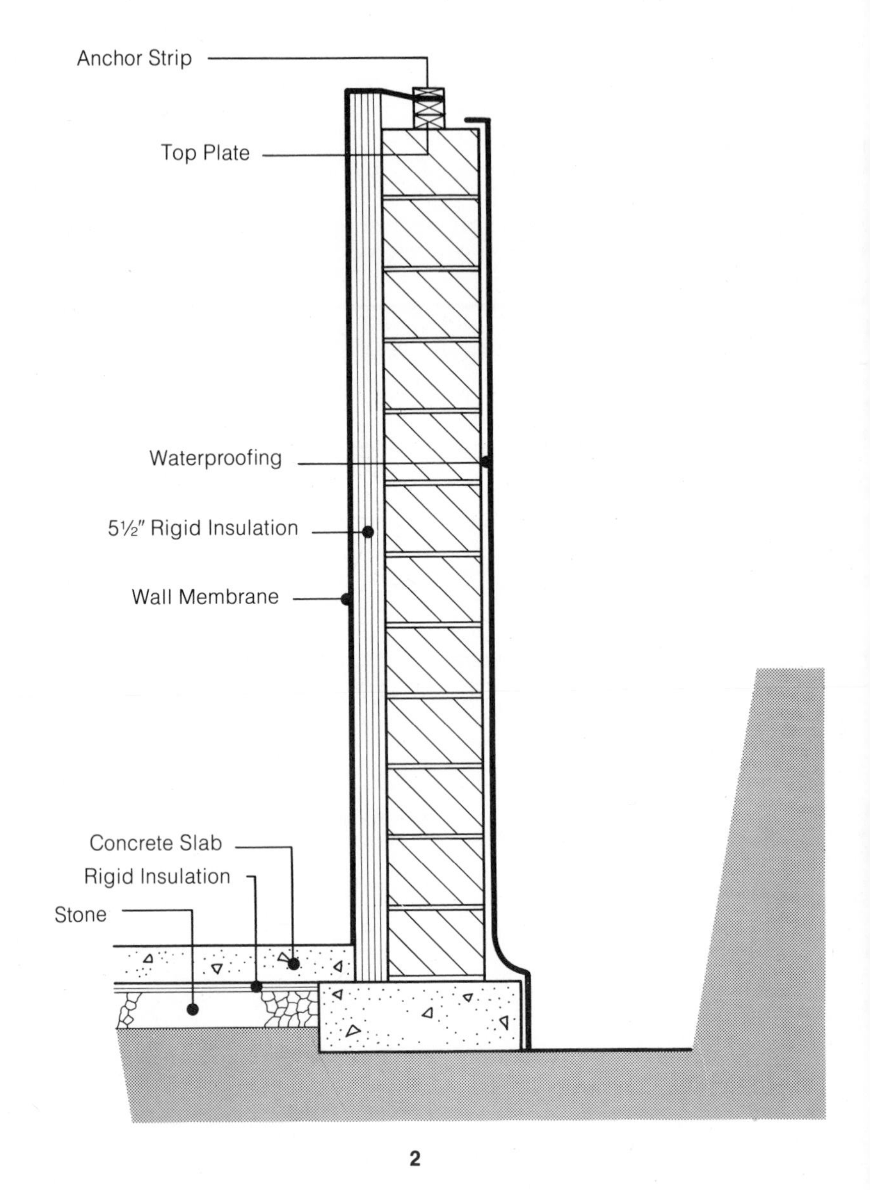

3–3. Stages of ESSPS wall construction.

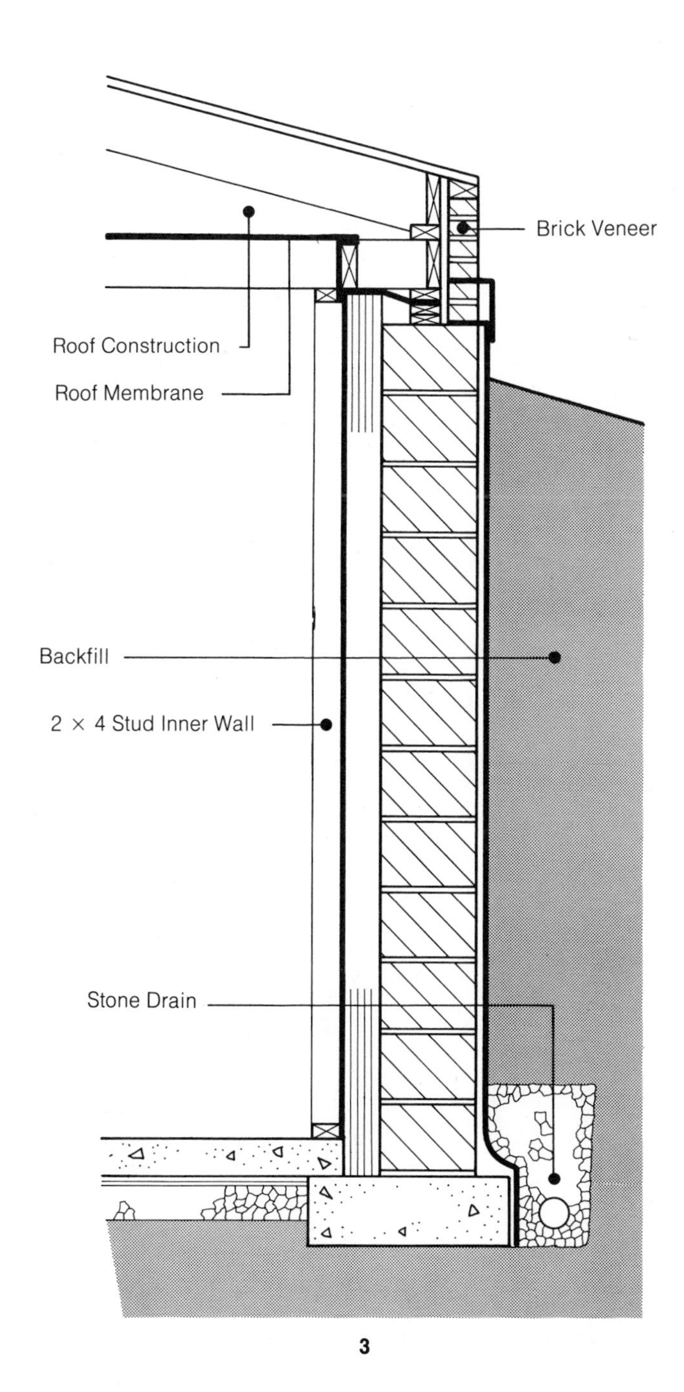
Brick Veneer
Roof Construction
Roof Membrane
Backfill
2 × 4 Stud Inner Wall
Stone Drain
3

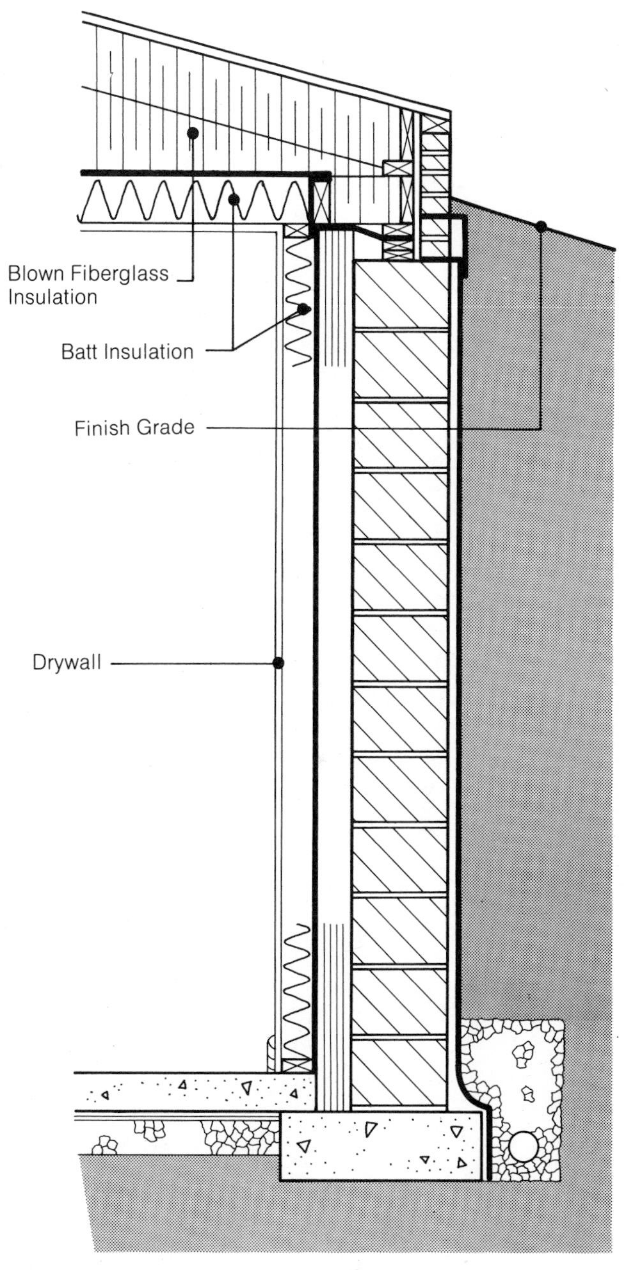
Blown Fiberglass
Insulation
Batt Insulation
Finish Grade
Drywall
4

5½-inch-thick EPS board.

The blocks are stacked dry, and mortar is replaced by surface bonding the block. A ⅛-inch-thick coat of cement is applied to both sides of the wall; glass fibers, embedded in the cement, hold the block together. The coating is also both airtight and watertight, so that multiple-layer wall construction can be reduced to a single-layer structure, built by a single trade. The wall coating can be finished to give the appearance of plaster or stucco, although extra labor is needed to achieve a satisfactory finish, and for interior work it is more economical to strip the walls and apply drywall.

Surface-bonded blocks do not resist earth pressure as well as conventional masonry walls, and use of reinforcing rods or other methods of strengthening the walls is required with earth sheltered design. Reinforcing requirements, combined with relatively high material and shipping costs for blocks manufactured in Canada, make the Sparfil system substantially more expensive than either the double- or triple-layer wall with structural shell of ordinary block. If conveniently located plants are opened in the United States, single-layer masonry construction of the ESSPS house, using insulating block, may become cost-effective.

The Roof

The A-roof on the ESSPS house will normally be built with 6-inch batts between the roof joists and a minimum of R-38 blown insulation on top of the membrane. With slab construction it is essential to place the membrane on top of the rafters, since not only electrical wiring but heating ducts and many of the plumbing runs should be located between the rafters rather than in or under the concrete floor, where they will be difficult to install and subject to deterioration from ground moisture. Placed on top of the rafters, the membrane will have very few joints (if the house is small only joints at the edges will be needed), and only plumbing stacks and ventilation ducts will penetrate it. The continuous roof and wall membranes, along with the protection provided by the earth shelter, create a virtually airtight building envelope.

Flat roofs can be used in place of the standard A-roof. The flat roof is visually very effective when used with a partially earth sheltered building. The roof will not be visible on the closed sides of the house, and the building can be designed so that its visible structure is formed of low walls that seem to rise out of the surrounding landscape. The closed sides of the house can also be hidden entirely with berms and low planting screens. But these design advantages are obtained by using a roof that, in comparison with the asphalt shingle roof laid over the A-framing, is costly, is not durable, and is likely to develop leaks. The design advantages of the flat roof must therefore be weighed against the economy and durability of the pitched roof.

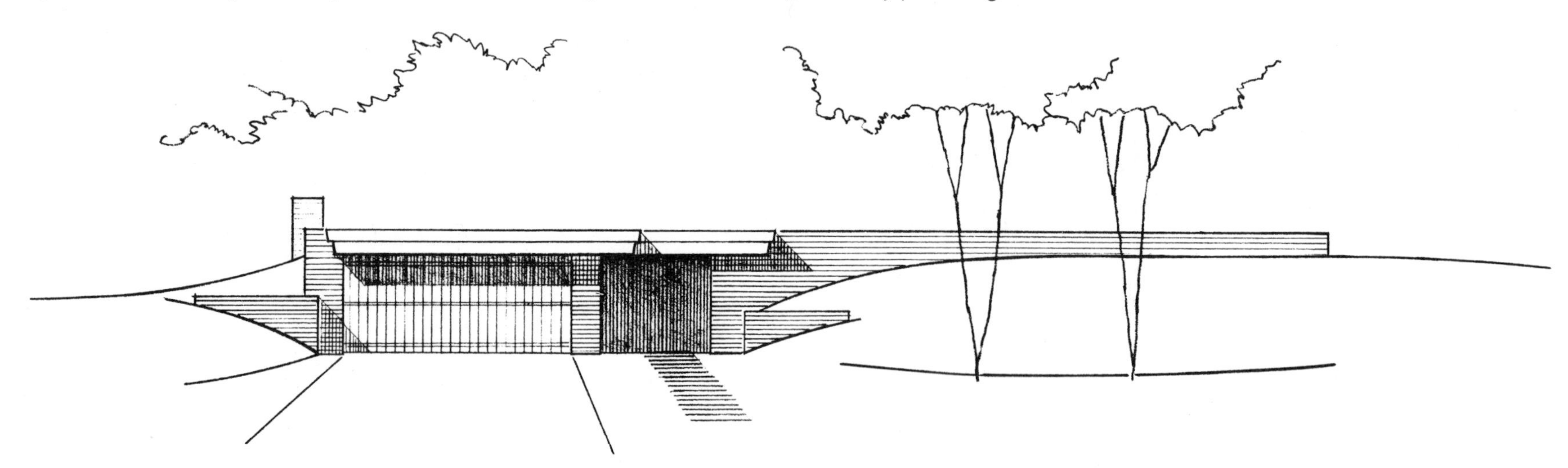

3–4. Elevation of partially earth sheltered house with flat roof.

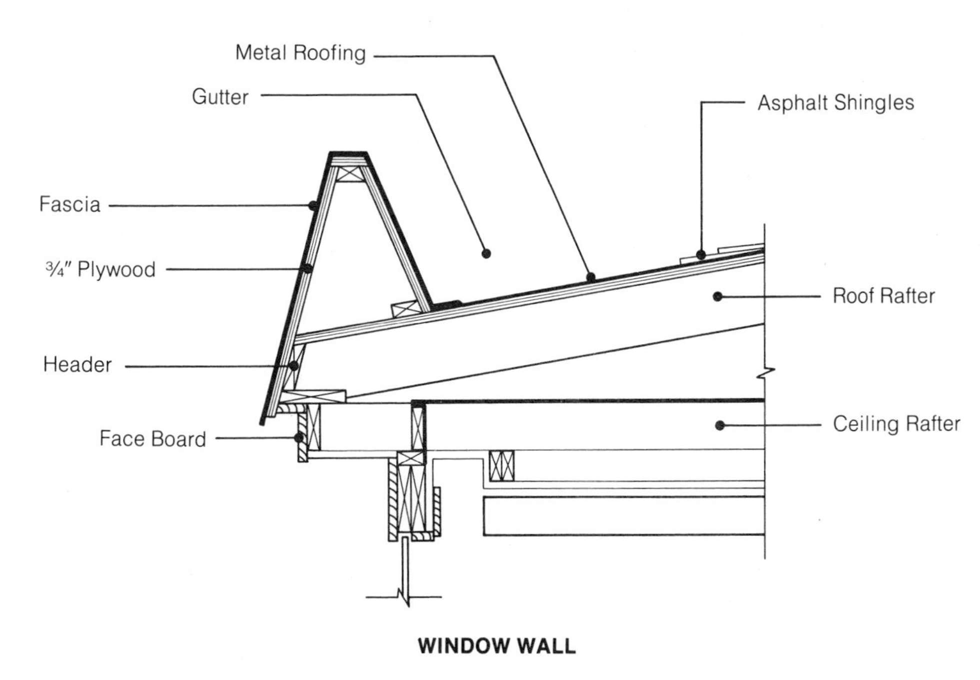

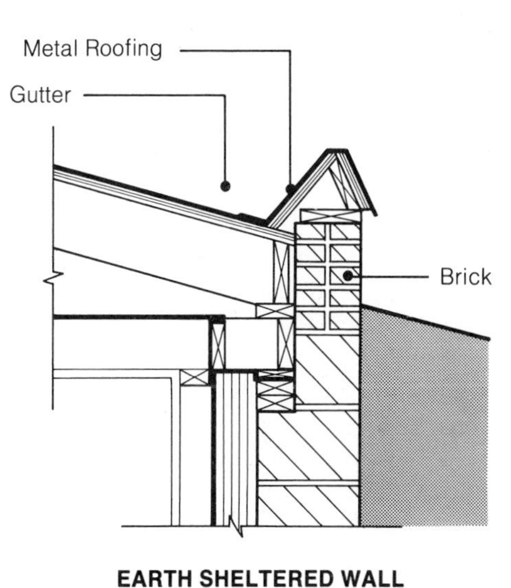

3–5. Falsework details for the A-roof.

Floor Construction

The concrete floor slab provides a large amount of passive solar storage capacity, and it is also much less expensive than a wood-framed floor built over a basement or crawl space. The major difficulty with slab construction is that the edges of the slab are exposed, and the building has cold, unevenly heated floors and a high rate of heat loss. Slabs with unprotected edges are also inefficient for passive storage. The ESSPS house solves these problems with earth shelter and multilayer walls. The slab is sunk 2½ to 8 feet into the ground at most points, and earth cover is backed up by rigid insulation, which extends down inside the masonry shell to cover the edges of the slab (see figure 3–2).

Earth cover and edge insulation should be supplemented by insulation placed under the slab. Adequate data on heat loss through concrete floors is not yet available, and the thickness and location of slab insulation remains a matter of the individual designer's judgment. Three to 4 inches of insulation should be placed under the slab where earth cover protecting the walls is less than 4 or 5 feet deep. This perimeter insulation must be at least 3 feet wide and in cold climates may be extended back from the wall as much as 6 feet. The interior of the slab can also be insulated, although the effect of this insulation is very difficult to estimate. Winter heat loss through an uninsulated inner slab will probably be very low, and in summer heat loss through the slab will help cool the house to a small extent. Conservative design would place a minimum (1 inch) of rigid insulation under the interior of the slab in climates with more than 5,000 heating degree days. This insulation should be omitted in more moderate regions, where the summer cooling effect will be more useful than the additional protection against winter heat loss.

Windows and Passive Solar Storage

Windows must be large enough to provide the house with adequate passive gain. Most of the glass in the ESSPS house should face south, but east- and west-facing windows can also be used. Local climatic conditions will determine the amount of glass that results in maximum cost-effectiveness for the passive solar system. In the Philadelphia metropolitan area optimum glass area for a medium-size (2,000 square feet of living area) house will be slightly under 400 square feet. This is equivalent to a continuous band of glass 4½ feet high running across the entire south wall of the house. Adding more than the optimum amount of glass reduces the cost-effectiveness of the design, because most of the extra passive gain will be "dumped" as waste heat to prevent overheating. Window area requirements over a wide range of temperature conditions, ranging up to 8,000 degree days with average sunlight availability, will generally be similar to those of the Philadelphia area house (see appendix E).

With passive solar storage requirements reduced by the heavy, airtight thermal building envelope, half the storage capacity of the passive system is provided by the building structure. The slab, thickened where appropriate, is the largest component of the storage system. Walls and ceilings of drywall or plaster also provide significant storage capacity. The remaining passive storage can be supplied either by internal nonstructural masonry walls or by tubes of phase change materials installed in a plenum attached to a central air-handling unit.

Storage capacity should be sized for limited temperature swings. Temperature swings in the ESSPS house do not need to exceed 8° F, allowing a temperature range from 67° F to 75° F. When the temperature inside reaches 75° F, extra heat acquired from passive gain will be dumped; this will occur on most clear and many partially clear days, even during cold months of the year.

Limited temperature swings restrict the ability of the passive system to store heat, but even with temperature swings held to 8° F to ensure comfort, only a relatively small amount of storage capacity must be provided by storage walls or phase change materials. The storage system is designed to heat the house without being replenished for a period of sixteen to thirty hours, during the three coldest months, and in late fall and early spring from forty-eight to eighty-four hours. Additional storage capacity that uses the energy discharged on clear days can easily be incorporated, but long-term storage capacity will not be cost-effective, because it will only be used during occasional periods of prolonged bad weather.

Efficient use of a passive solar system requires adequate thermal protection for windows and skylights during the long hours of winter darkness. In the Philadelphia metropolitan area, heat loss through unprotected double-glazed south-facing windows will be over 90 percent of their daytime solar gain. In more severe climates, unprotected south-facing windows will often lose more energy than they gain. Thermal shutters, practical in the ESSPS house because its window area is typically only about 25 percent larger than that of the conventional house, provide the best solution to heat loss through windows. Shutters can be built to provide very high thermal resistance, since adding extra insulation does little to increase their cost. R-17 (3½-inch-thick) shutters are appropriate for climates with up to 6,000 degree days. R-22 (4½-inch-thick) shutters should be installed for more severe winter conditions. The shutters will allow more than 50 percent of the passive gain to be retained, so each square foot of glass area provides a far larger net gain than if the windows are left lightly protected or are entirely unshielded.

Windows and skylights must also be protected against direct solar radiation during the summer. Direct sunlight can be blocked with lightweight, removable overhead sunscreens that are put up in the late spring; the bermed design of the ESSPS house makes this a simple task. Windows should also be equipped with venetian blinds or drapes that provide protection against indirect summer solar radiation, supplementing protection against direct radiation provided by the overhead screens.

Mechanical Equipment

The ESSPS house has a backup heating system as well as a heat exchanger, fans that help remove internal heat gain in summer, and a hot-water preheat system. Backup heating is used infrequently, but the system must be capable of adequately heating the house in very cold weather. The capacity of the backup system will be half that of the heating system of a conventional house, and almost any type of heating equipment, including fireplaces and wood-burning stoves, can be used. Most homeowners will, however, prefer the convenience of a conventional central heating system. Air-conditioning equipment will also frequently be installed, since residential air-conditioning is regarded as a necessity in most parts of the United States. The smaller cooling loads of the ESSPS house will not keep temperatures down inside the house, but they will reduce the size of the air-conditioning equipment and the amount of energy consumption required for cooling.

While the backup heating system operates infrequently, substantial amounts of fresh air must be drawn into the tightly sealed ESSPS house at regular intervals to maintain habitability. Ventilating air is drawn into the house through the heat exchanger, allowing the heat of the exhaust air to be recovered. Current generation air-to-air heat exchangers are capable of recovering approximately 80 percent of the heat of the exhaust air, although part of this savings is offset by the cost of operating the exchanger, which uses a substantial amount of electricity to run its intake and exhaust fans.

The hot-water preheat system uses part of the surplus passive solar gain that is available throughout the heating season. Cold water enters the hot-water heating system at approximately 50° F. Air warmed by passive gain is used to raise the water to a temperature that is close to the maximum room air temperature of 75° F. If the hot-water discharge is set at 110° F, preheating will save 40 percent of the cost of hot-water heating, and the system will also reduce energy needs for summer cooling. In hot weather the preheat system also helps to cool the house as it absorbs heat from the room air.

4

Design Characteristics of the Combined Approach

Organization of the Plan

The use of passive solar, combined with partial earth shelter, dictates a house plan with living spaces located on the south side of the building and service spaces, including the garage and entrance, on the sheltered north side. A great deal of flexibility in orientation and plan arrangement is possible within this framework. The orientation of major living spaces and glass areas can be turned as much as 25 degrees from true south, so that all or part of the building can face southeast or southwest, with little effect on the overall performance of the passive system. In addition, windows can be placed in the east and west walls of the house. They will acquire smaller passive input here than the south-facing windows, but placing as much as 15 to 20 percent of the glass in the end walls will, again, have little effect on overall passive performance.

The division of living spaces within the ESSPS house can be made on the basis of planning and aesthetic considerations with relatively few limitations imposed by the structure. The ESSPS house is well adapted to the open plan that treats kitchen, dining room, and the living room as one large space. This space can be left almost entirely open or, if the designer wishes, room dividers can be used to form subspaces (figures 4–1, 4–2, 4–3). The ESSPS house can also be planned in the same manner as the conventional tract house. The medium-size tract house usually has separate spaces for sitting or recreation, eating and food preparation. An ESSPS house that uses this conventional arrangement of the living spaces will be strongly linear (figure 4–5).

The central corridor that connects the multiple living spaces of the linear plan can readily be turned into a wide skylit gallery by installing several inexpensive and virtually leakproof plastic skylights. Widening the corridor under the skylights creates a gallery that can be filled with plants growing out of simple openings in the floor slab or set in raised, decorative planters. If the budget permits, the decorative effect can be enhanced by laying ceramic tile over the slab along the pathway through the plants. The extra space needed to turn the corridor into a skylit atrium-gallery is comparatively inexpensive, since it is located in the center of the building.

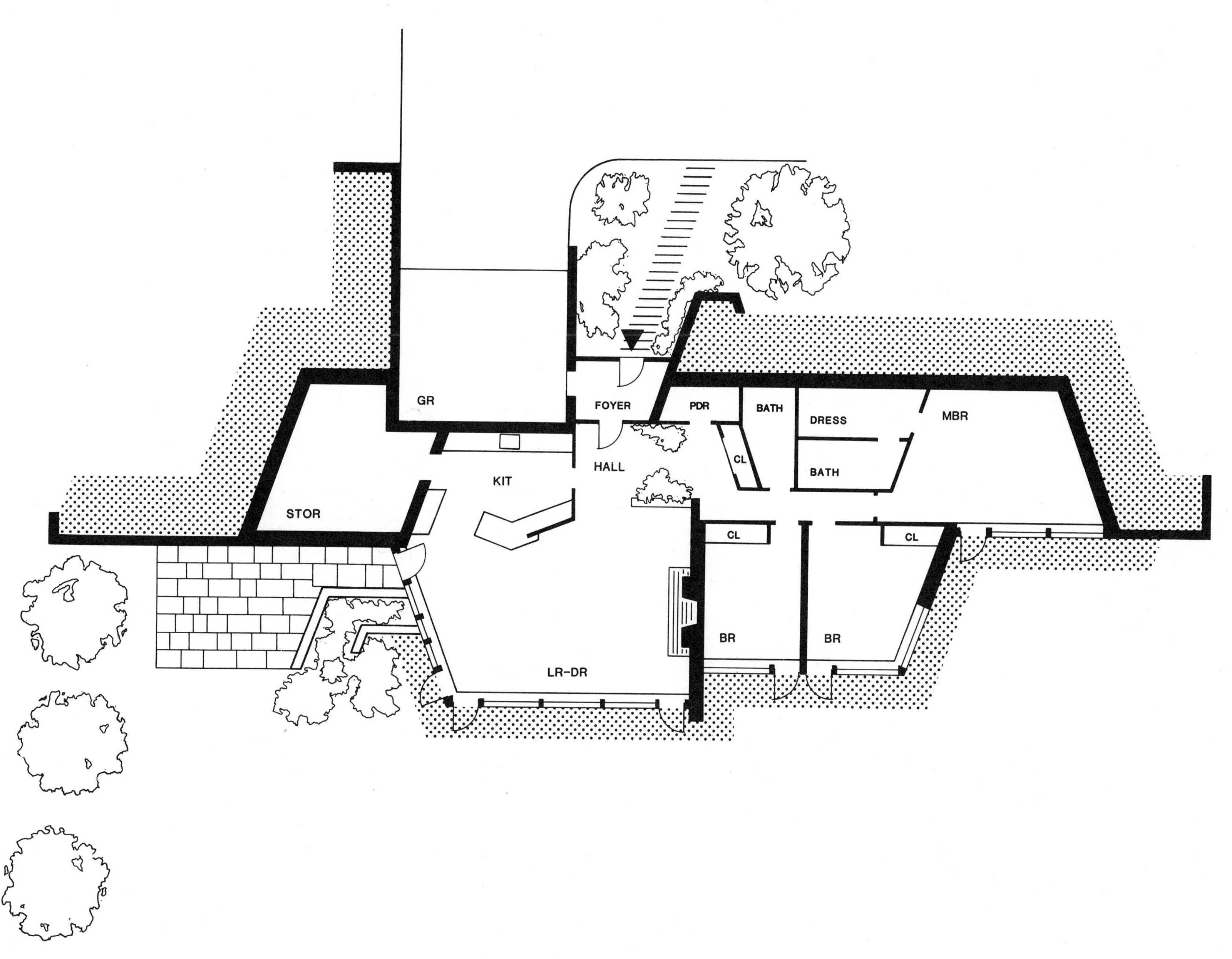

4–1. ESSPS house plan with open living spaces.

4–2. Garden view of house with open plan.

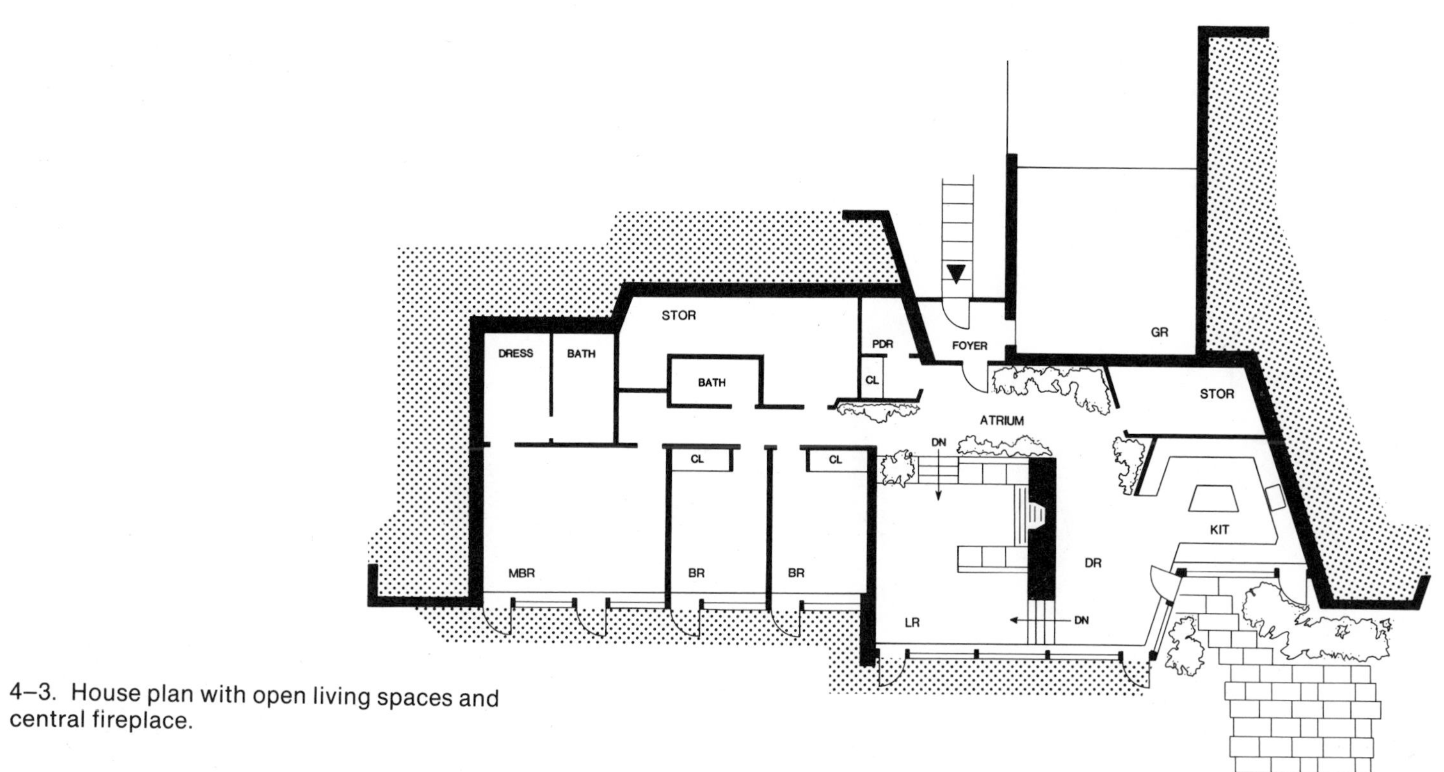

4–3. House plan with open living spaces and central fireplace.

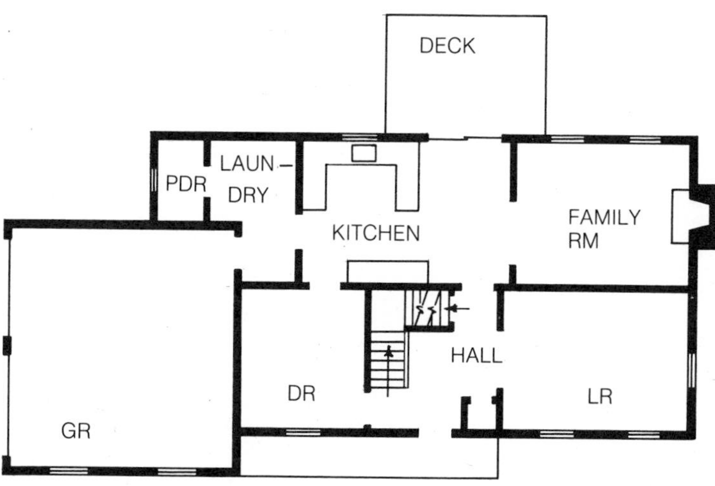

4–4A. Conventional tract house.

4–4B

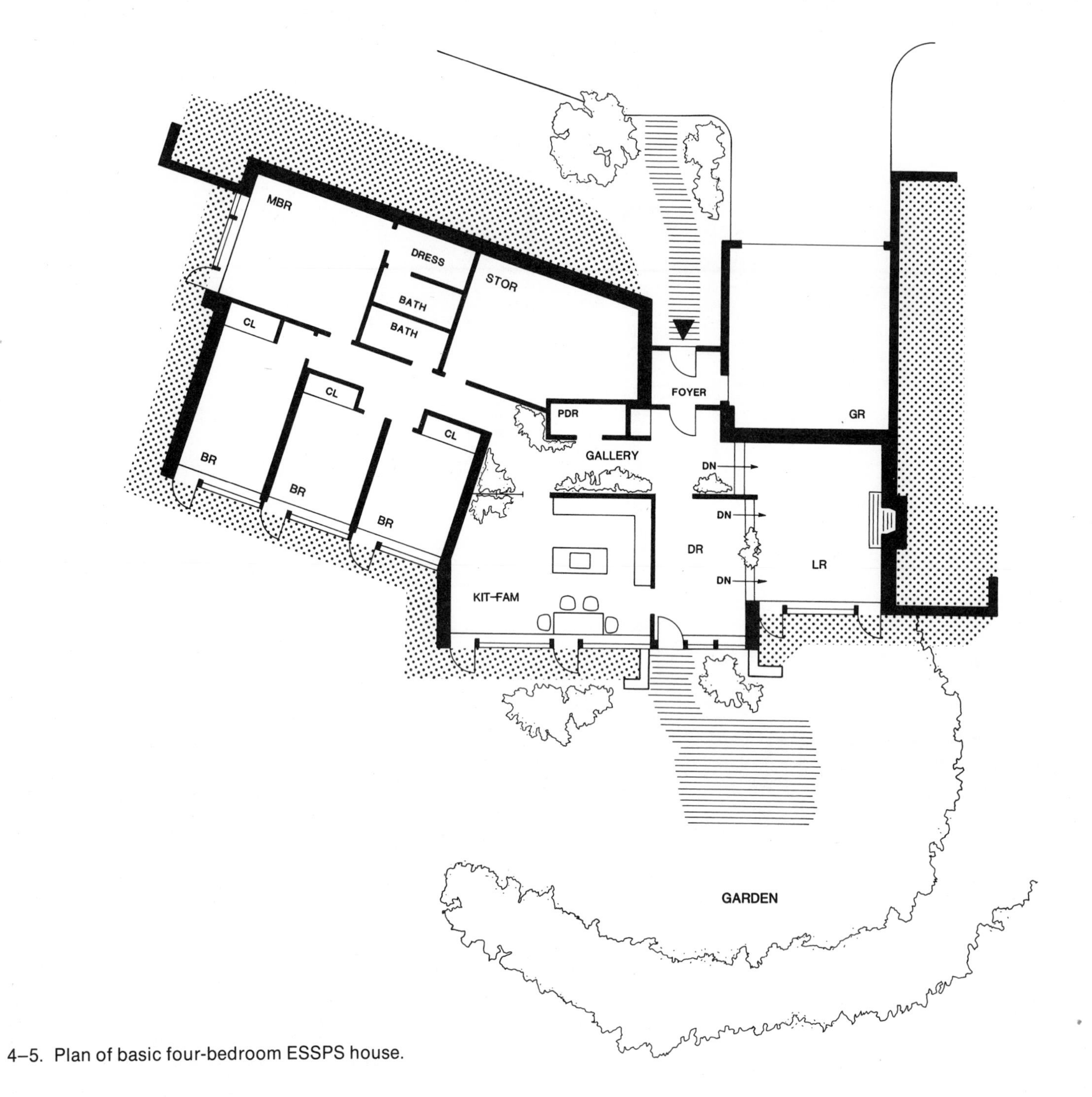

4–5. Plan of basic four-bedroom ESSPS house.

The central corridor can be articulated by varying its width, introducing changes in level, and using angled wings. The angled wing introduces a change in the direction of the corridor and reduces its length by several feet, although overall building length does not change (figure 4–5). L- and U-shaped plans, in which one or more rooms face east and west, will shorten both building and corridor length (figure 4–7).

The wide central corridor leads to the front entrance, which is next to the garage. The entrance side of the ESSPS house is protected by berms, so separating the front entrance and garage with a traditional "back" door is not practical. Air locks, which are simply old-fashioned vestibules, are needed on the entrance side to minimize heat loss when the doors are opened, and eliminating the back door reduces construction costs; two entrances require two air locks and two pairs of expensive exterior doors. If the main entrance is next to the garage, only one air lock is needed and a total of three exterior entrance doors rather than four is required.

4–6. Garden view of basic ESSPS house.

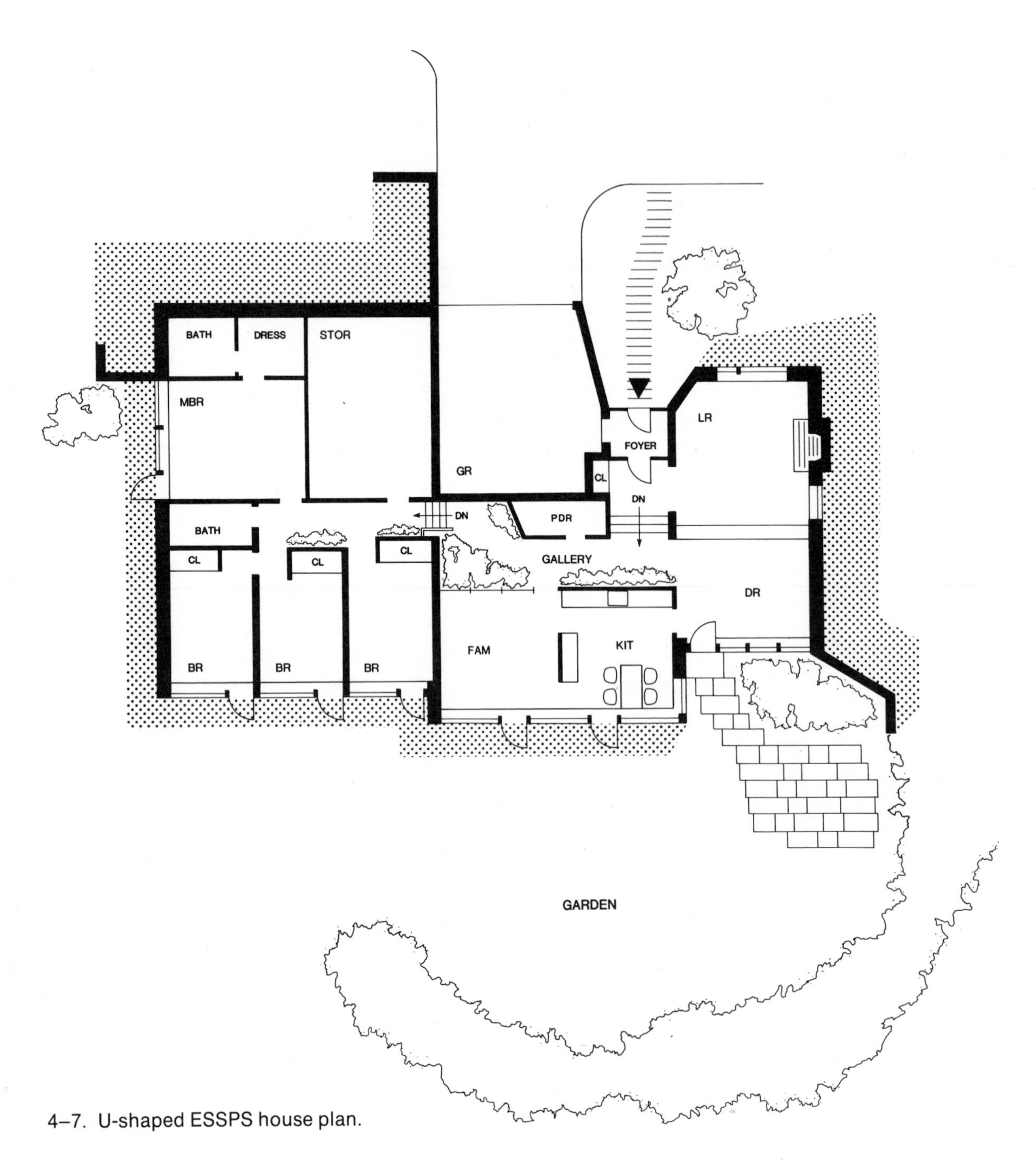

4–7. U-shaped ESSPS house plan.

4–8. Entrance side of the U-shaped house.
Garden view of the house with arched windows.

Two-story Construction

The floor slab of the one-story ESSPS house provides a large amount of inexpensive and well-distributed passive storage capacity. Equally important, its one-story multilayer wall is extremely simple to build. Adding a second floor will make wall construction more complicated and also increase the number of membrane joints. The two-story plan does save some money, because it has a smaller circulation area and a smaller roof. In addition, the combined area of the exterior walls and roof is slightly reduced, creating a small savings in heat loss from the building envelope. These advantages are more than offset by complicated construction and reduced slab storage.

Frequently, however, restricted lot size or design considerations will necessitate two-story construction of part or all of the house. The two-story house must be planned so that high earth sheltered walls are not required. Two-story earth sheltered walls are very difficult to build; they can only be constructed if the house is cut deeply into a steeply sloping natural grade. Even on a sloping site the high walls will require very heavy construction and will be very expensive (figure 4–9A). The chances of water leaking through the walls are also much greater than in a one-story structure. On flatter sites it will be virtually impossible to build two-story earth sheltered walls, since the closed sides of the house will be covered by high berms, and massive construction is needed to resist the pressure of the fill. The berms themselves, raised to a height of 16 or 17 feet, will usually be visually awkward.

In place of high earth sheltered walls, the two-story ESSPS house can use stepped walls. The part of the structure built against

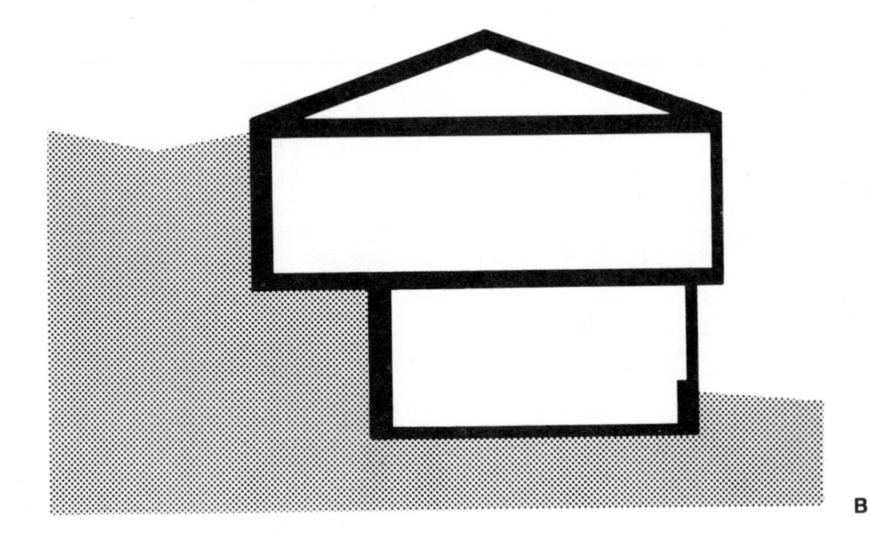

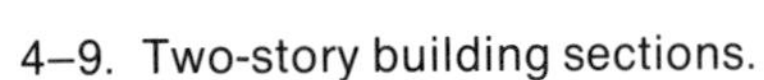
4–9. Two-story building sections.

the earth is made up of two separate one-story sections, so the extra heavy construction of the straight two-story wall is not needed (figure 4–9B). This design is generally restricted to south-facing hillsides, where the building can be set into the natural grade. A second and more flexible two-story ESSPS structure combines a one-story north-side service area with south-side spaces arranged on two floors. This structure can be built with earth sheltered walls protected by artificial berms, so that it is not subjected to the site limitations of the stepped design (figure 4–10).

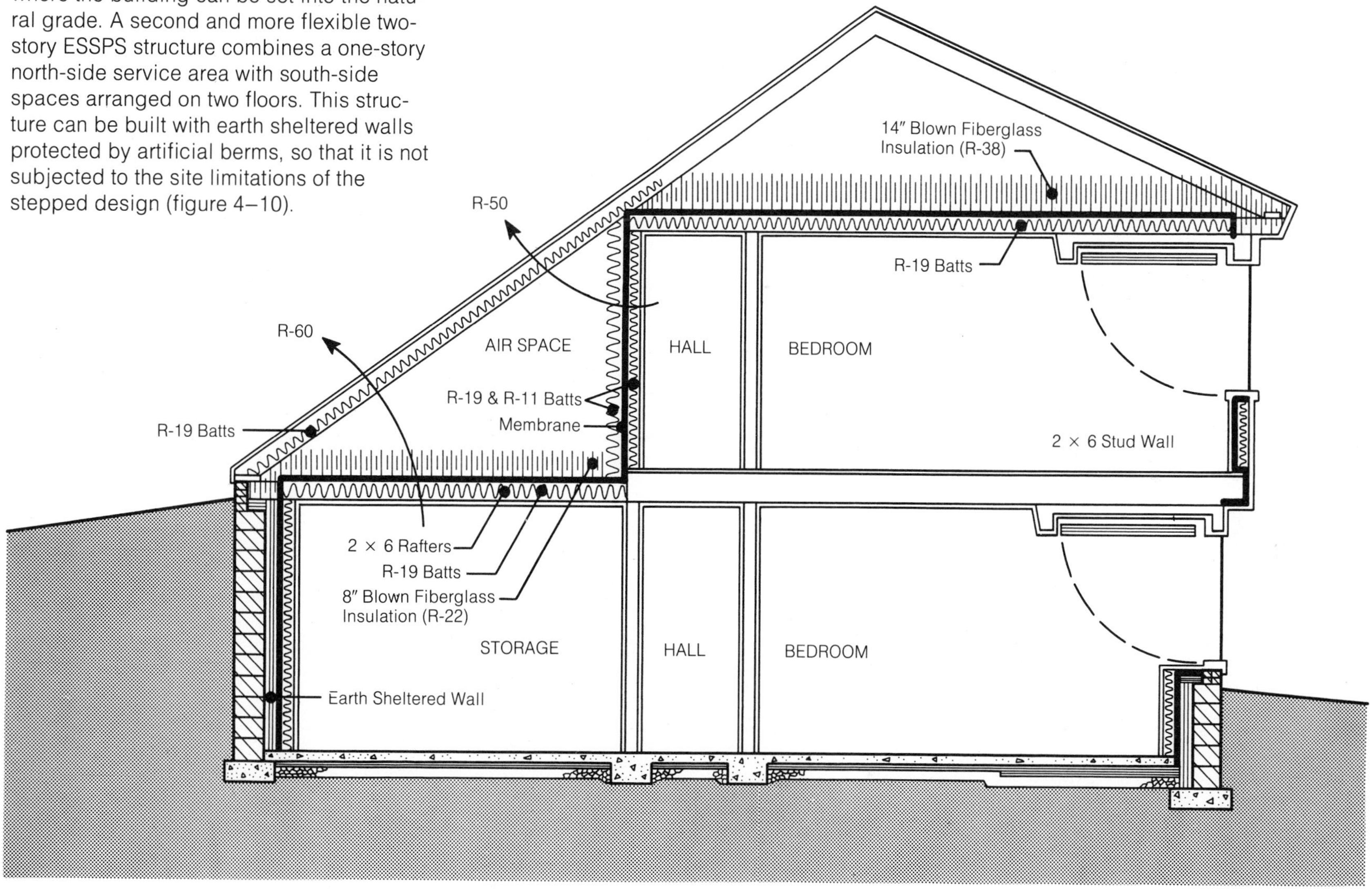

4–10. Cross section of two-story ESSPS house.

Elaborating the Plan

The concrete floor slab and block outer walls of the ESSPS house can be constructed without difficulty in almost any shape. One or two angled walls can be introduced into a rectangular plan, or the entire house can be developed on a hexagonal or octagonal grid. Angled walls permit the placement of major spaces around a central atrium, eliminating the central corridor and creating a winter garden in the heart of the house (figure 4–11). Curved walls can also be introduced, if the arcs are not too short; this allows the living spaces to be arranged in a segment of a circle, somewhat on the pattern of Frank Lloyd Wright's famous Herbert Jacobs House II (figure 4–14). Block for the curved walls can be laid up without difficulty, and the stud inner shell can be laid out in straight segments 8 to 10 feet long.

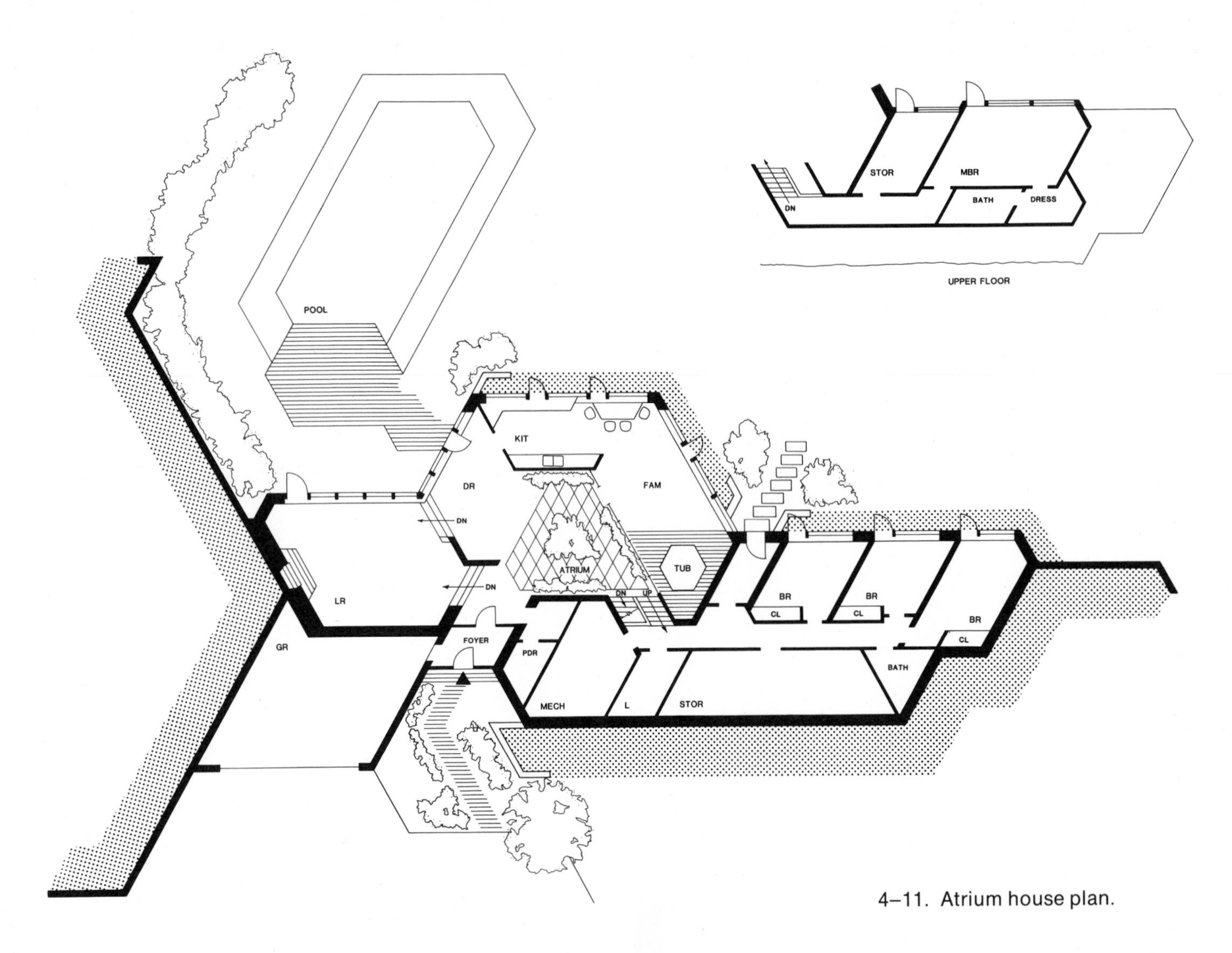

4–11. Atrium house plan.

4–12. The atrium.

4–13. The living room in an atrium house.

Flat roofs can easily be framed to fit nonrectangular layouts, but sloping roofs will require numerous hips and valleys. Rafters and roof sheathing must also be cut to changing patterns, slowing the work. This represents the only major construction complication encountered with nonrectangular forms. The rectangular house will always be the most economical to build, but the extra costs attached to complex and varied plan forms are much smaller than those encountered with conventional frame construction.

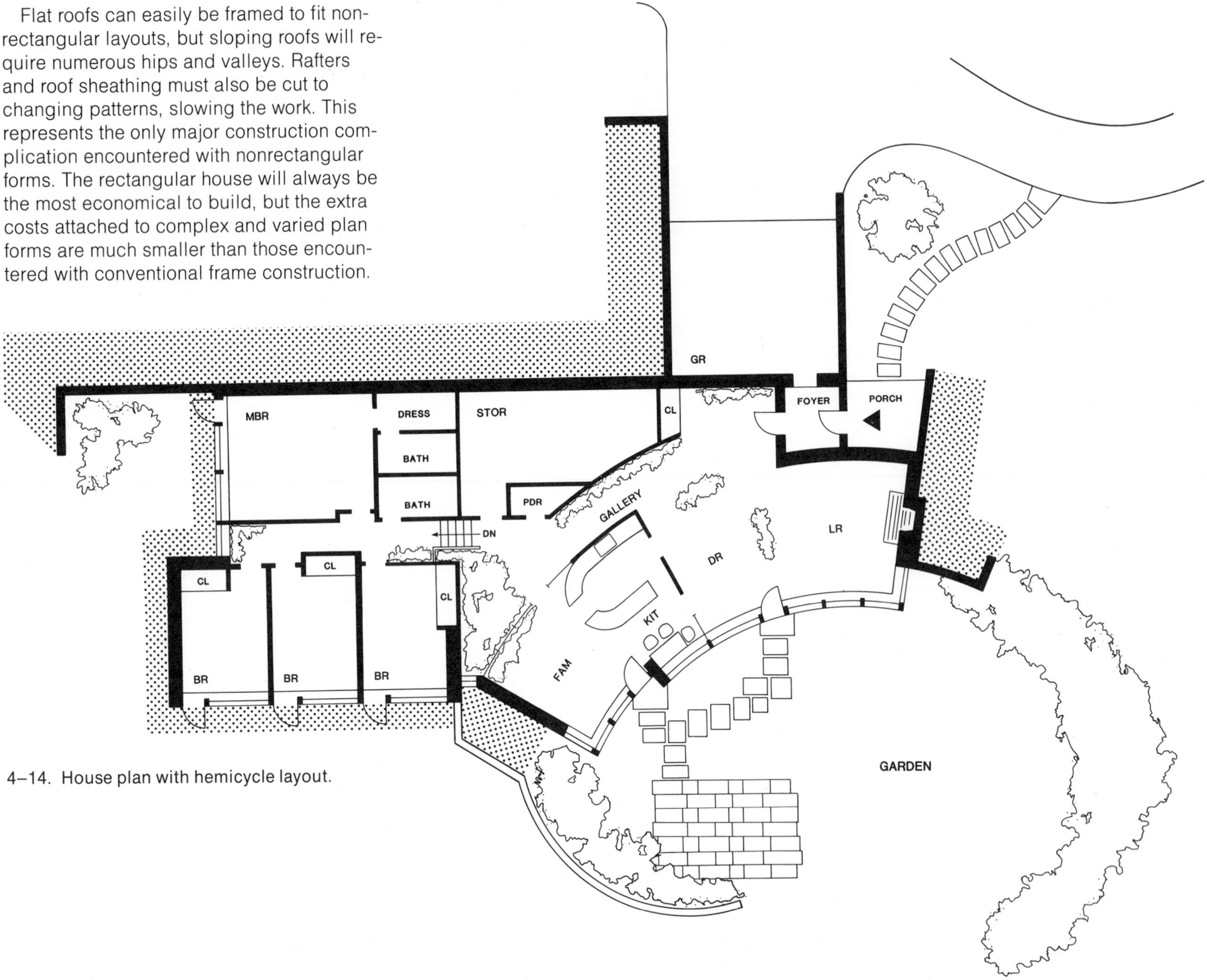

4–14. House plan with hemicycle layout.

4–15. Entrance and garden views in hemicycle house.

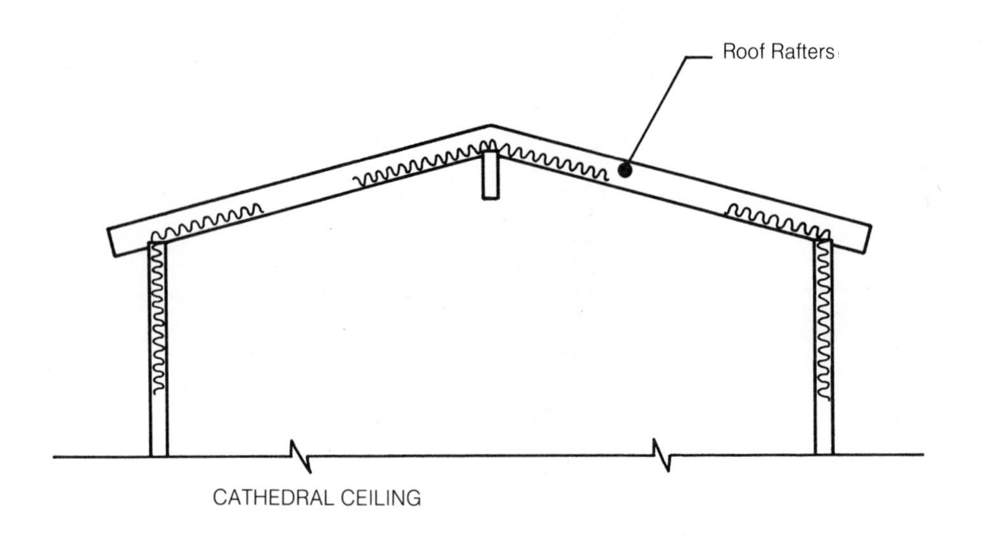

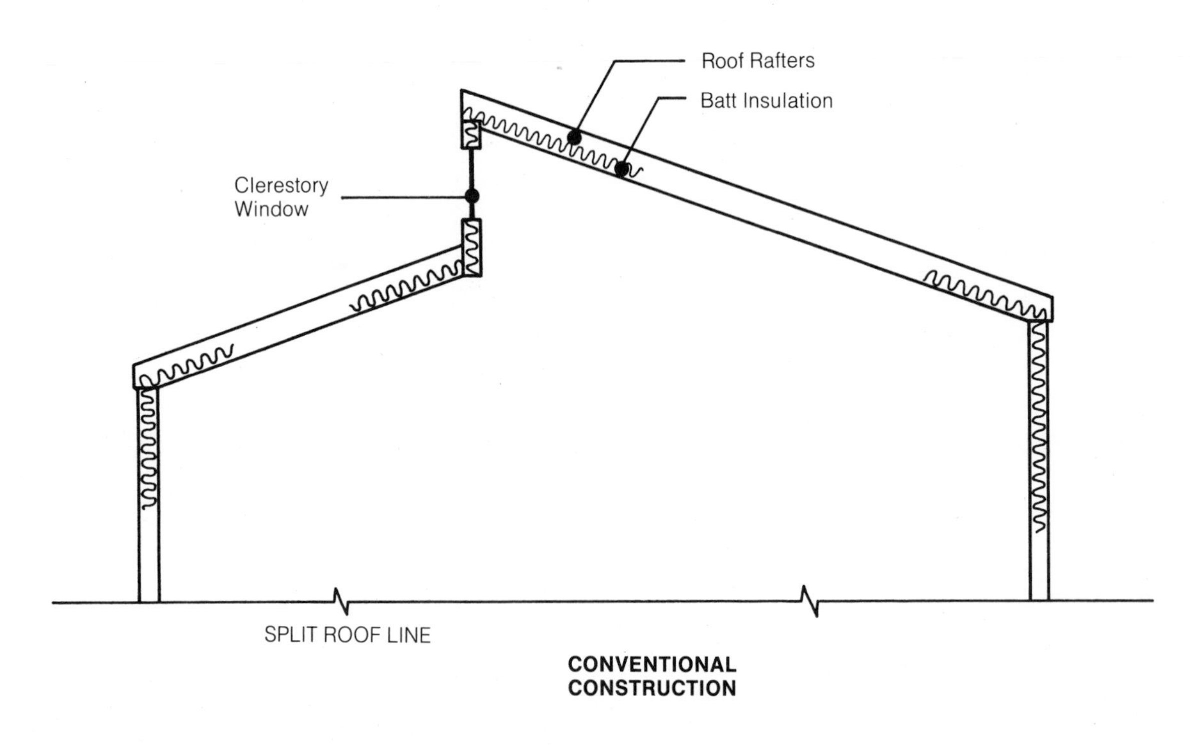

4–16. Framing high spaces.

Development of Interior Space

The treatment of the interior space of the ESSPS house should take account of construction limitations imposed by the heavy, airtight thermal envelope, while taking advantage of the architectural possibilities of earth sheltered design. Elaborate vertical development of the interior is difficult, because sloping ceilings complicate the roof framing. Conventional cathedral ceilings are usually framed with a single set of roof rafters (figure 4–16). Superinsulation requires a deeper roof and as a result the superinsulated cathedral ceiling or sloped, split roof is awkward to construct. Wherever possible the upper space of the ESSPS house should be elaborated by building down within a simple envelope. This can be done by adding decks, lighting coves, and stepped ceilings within the volume of a rectangular envelope (figure 4–17).

Spatial variation can be achieved with changes in floor level, which are introduced into a house built on a slab simply by pouring a few steps between adjacent rooms. The changes in level may include steps leading up or down from the entrance, steps between the living and dining areas, and "conversation pits" in the living and family rooms. On many sites it will also be necessary to vary the slab elevation in order to follow the grade. Here there will often be steps between the living spaces and a bedroom wing.

If abrupt changes in ceiling height and floor elevation are avoided, there will be strong emphasis on horizontal spatial movement. Emphasis on horizontal spatial development permits logical construction and is an appropriate visual design for a structure with earth sheltered walls and an extended plan. Horizontal spatial development is also appropriate for the multiple living-space arrangement of the conventional tract house. Walls between these spaces can be partially or wholly eliminated without creating structural problems. Visual emphasis will shift from the living spaces to the areas that surround them—the garden and the skylit central spine or atrium. The individual spaces will be open to each other and to the central spine, and plants can be used as part of the screen walls that separate the spaces, allowing greenery to dominate the interior.

Exterior Space

The conventional house uses its site as a surface upon which the structure is raised. Landscaping and site design create a frame around the building, defining the surroundings against which the structure is visible. Landscape and building blend in earth sheltered architecture, and the visible elements of the partially earth sheltered house are a combination of building structure and shaped earth forms. The earth forms join the house to the surrounding land, and the boundaries between landscape and architectural elements disappear.

This characteristic of earth sheltered design is clearly visible on the street or entrance side of the partially earth sheltered house. The approach to the house is through a courtyard, formed by the berms that protect the house. The visible portions of the house may be restricted on this side to the garage and entrance, with the rest of the house concealed by banks or berms, retaining walls, and plantings. More of the house can be seen if the retaining walls hold back the earth berms to reveal an entry facade. Architectural conventions that clearly indicate the earth sheltered character of the house may be used for the facade, or, if the client or builder desires, the facade can be designed to resemble that of a more conventional building.

The courtyard itself can become a tightly

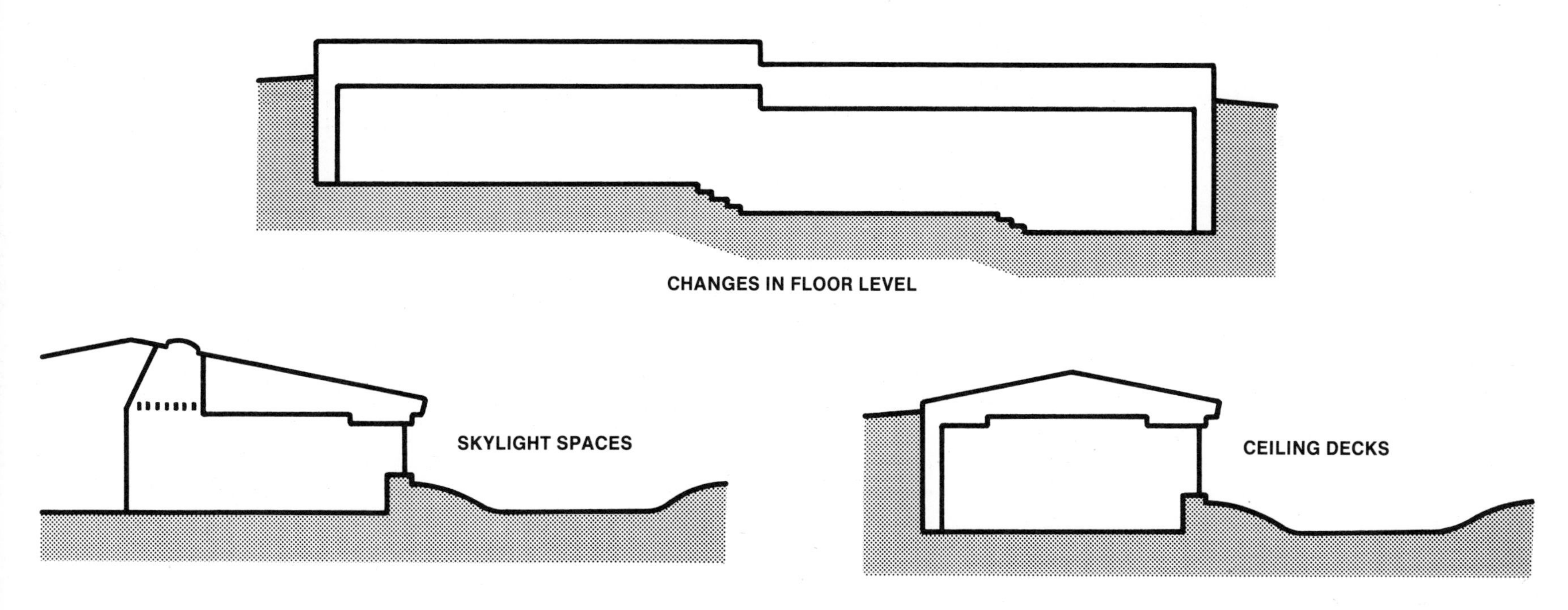

4–17. Spatial development of the ESSPS house.

enclosed area or remain a relatively open space. On steep slopes, grades may restrict its size, but on flatter ground there will be greater freedom to shape the courtyard with earth forms and planting screens. The approach side of the courtyard can remain open, leaving the house visible from the street or road, or it can be partially or wholly enclosed by extending the berms from the house.

The garden side of the house offers a second, even more important, opportunity to join exterior and interior space. Here the house, set deeply in the ground, opens onto its garden through a wall that is largely glass. Extending the berms that cover the ends of the building forms an enclosed garden with a windbreak that protects the house at the same time that it defines another exterior space. This completes a spatial sequence that begins in the entrance courtyard, moves through the interior of the house, and extends out through the broad windows of the south wall into the sheltered garden.

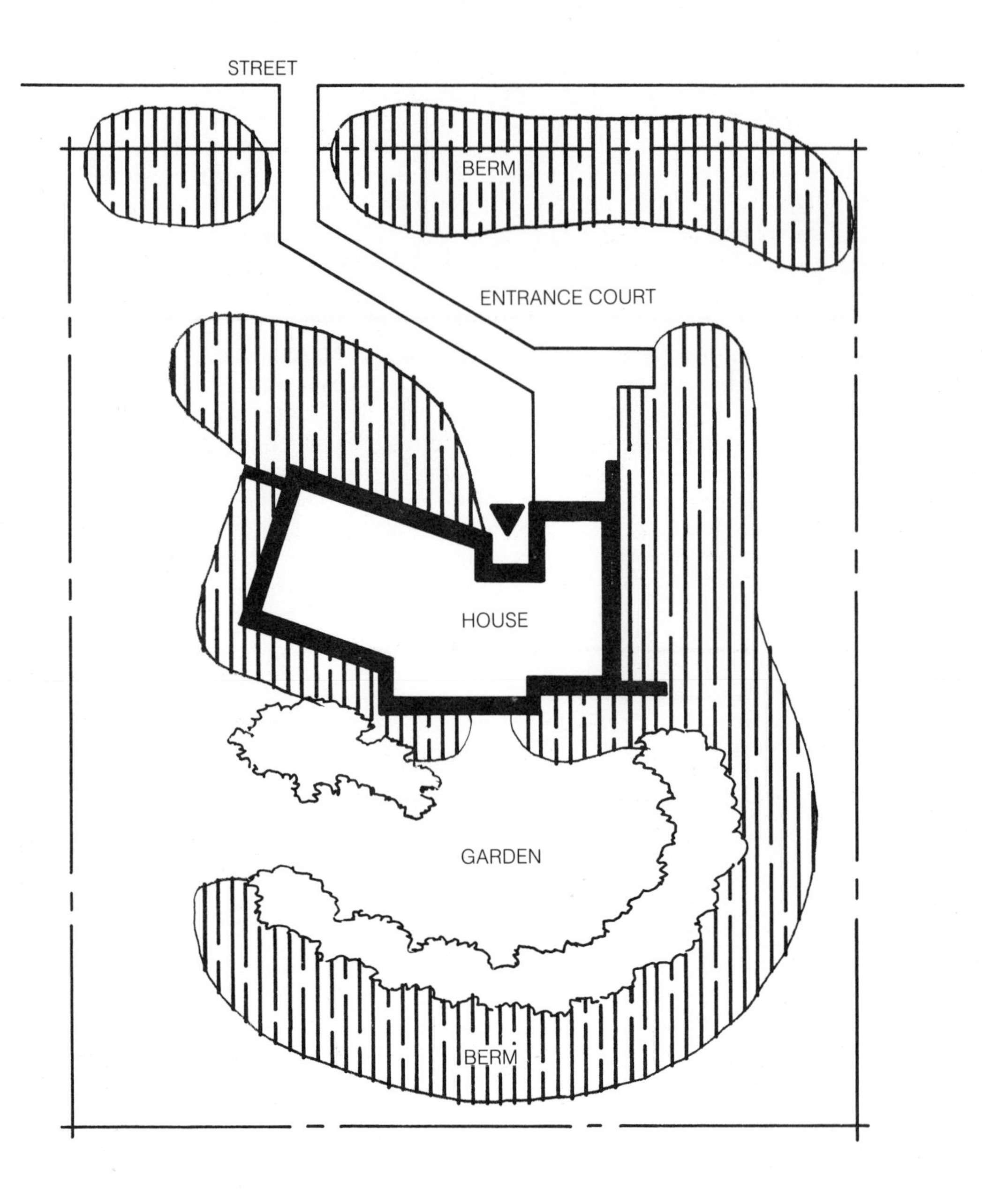

4–18. Lot design of the partially earth sheltered house.

PART II

MATERIALS, STRUCTURE, AND CONSTRUCTION DETAILS

5

Materials and Construction Methods

The ESSPS house is built with familiar materials and construction methods. The floor of the house is an ordinary concrete slab, the outer walls are masonry or concrete, and the inner walls, partitions, and roof use frame construction. The high level of thermal protection and resistance to air infiltration is achieved with fiberglass batt and blown insulation, polystyrene board, and polyethylene sheets—materials that have been widely available for years, and with which the home-building industry is thoroughly familiar. Mechanical equipment is simple and does not require complex controls. But while the basics—materials, construction, and mechanical equipment—are not new, the efficient combination of earth shelter, passive solar, and superinsulation requires innovation in the design of many parts of the building, as well as careful analysis of construction procedures.

Materials Selection

Selecting the exterior materials for the ESSPS house presents few problems. Masonry construction is used for the earth sheltered walls, and where the walls emerge from the earth this construction is continued, forming an all-masonry exterior. Exposed exterior walls can be left unfinished if they are concrete; block walls can be finished, above grade, with stucco or brick. Stucco, directly applied to the structural walls of the house, forms a durable and attractive finish, but for most earth sheltered designs brick is the ideal material. The all-brick exterior of the medium-size partially earth sheltered house requires only 500 or 600 square feet of brickwork, and a brick exterior can be included within the limits of a comparatively modest budget.

In addition to its masonry walls the ESSPS house has wood trim for windows, doors, bargeboards, and soffits. Cedar, redwood, or similar materials should be used to ensure maximum durability and reduce maintenance costs. Woods such as cedar and redwood can be stained, rather than painted, and restaining is much simpler and less expensive than repainting trim. All trim on the one-story ESSPS house can be reached from the ground, with the occasional need for a short step ladder, and periodic restaining of the trim can easily be done by the homeowner.

Interior materials will generally be similar to those of a conventional home. The less expensive ESSPS house will have walls and ceilings of gypsum board (drywall); a more elaborate design will make extensive use of wood or other decorative materials. The masonry walls introduced for passive solar storage add variety to the interior, particularly when gypsum board is used for most of the other interior surfaces. Storage walls of a house built on a limited budget are of exposed cement block. Solid block, or filled hollow blocks, will maximize storage capacity of the walls. The storage walls are nonbearing and can be laid in a stack bond pattern and then painted bright colors. The construction budget should allow for some extra care in constructing the walls, including rubbing down the joints with a stone. Stucco or plaster finish provides an alternative to painted block, although plastering the block will nearly double the cost of the walls. Where the budget permits, brick can be used, making the walls a handsome addition to the interior of the house.

An exposed floor slab can provide nearly one-half the total required passive storage capacity. But bare concrete, even if integrally colored, will generally be unacceptable in the living areas of the home. With a sufficiently generous budget ceramic tile, which combines a handsome appearance with excellent heat transmission properties, can be laid over the slab. Ceramic tile floors, however, are extremely expensive, and it is more cost-effective to use less expensive materials, such as vinyl sheeting and carpet. Carpet will reduce the storage capacity of the slab, because it slows the rate of heat transmission from room air to the concrete below the carpet; vinyl will have only a limited effect on heat transmission and storage capacity. If the bedrooms and the living room are carpeted, 25 percent of the total storage capacity of the slab will be lost. The savings in the cost of the flooring in comparison to materials such as ceramic tile will, however, be many times greater than the cost of the extra storage capacity needed to make up for the reduced slab storage.

Windows are a major architectural feature of the ESSPS house, and special attention should be paid to the window trim. Stained trim can be used to create a strong decorative element in a house with simple interior finishes and will also enhance the effect of the window wall. Where trim is stained, mahogany can be used in place of pine; good mahogany trim costs less than comparable grades of pine. More time will be needed for the carpentry work with stained trim, because joints must be fitted more accurately. The interior painting also requires greater care to avoid splashing paint or stain on the contrasting materials, but if the trim is prestained, a large amount of extra labor will not be needed. The total additional cost of the stained mahogany trim in a medium-size house will be in the range of $300 to $500.

Construction Procedures

The structure of the ESSPS house combines masonry and frame construction. The initial stages of construction, during which large quantities of heavy materials including block, concrete, and crushed stone are brought to the site, are particularly important, and efficient handling of materials at this stage of construction is necessary for controlling the cost of the building operation.

Block trucks, with their long booms, should be able to place their loads immediately adjacent to the masons' work stations; the stone that goes under the slab should be unloaded inside the house, and concrete slabs should be poured directly from the mixing trucks, without using wheelbarrows or pumps. Once the structural walls and the slab are completed and framing begins, the lumber should be unloaded just outside the house, so carpenters do not have to carry the lumber piece by piece through the mud for a distance of 30 or 40 feet.

Constructing the shell of the ESSPS house can be divided into eight steps:

1. Grading the slab, and digging the footings.
2. Pouring the footings.
3. Erecting block or concrete structural walls.
4. Pouring the slab.
5. Erecting the frame inner shell and interior partitions.
6. Roof framing.
7. Rough grading berms.
8. Completing masonry work.

When the building site is graded for the floor slab, the bulldozer or front end loader should also level the ground 40 to 50 feet around the house, allowing heavily loaded trucks enough room to approach the building from all sides.

On a sloping site, access will often be limited to the south side of the building. Earth taken from the hillside as the site is leveled should be placed at least 50 feet from the building. Piled here, it will later form the garden berm, while a wide apron for maneuvering large trucks is left in front of the building. Providing access for the materials trucks on a hillside site will usually require rough grading a temporary drive on the south side of the lot.

Footings can be poured directly from the concrete trucks. The trucks can approach the building from all sides if the lot is flat. On a sloping site, some of the footings will be poured from trucks that have backed into the area leveled for the slab. This can be done easily, as footings on the south side are always dug at grade, and there are no deep trenches to cross. The same procedures can be followed when the block is unloaded or when pouring concrete for the structural walls, so that all materials for the structural walls and footings can be brought directly to where they are used (figure 5–1).

After the walls are up, the stone base for the slab is laid down, and the slab is poured, but access to the house is restricted; only the garage and the south wall remain partially open. Stone trucks can back into the garage area, on the north side of the building, or dump the crushed stone for the slab over the low walls on the south side of the house. The slab is large enough to justify leveling the stone with a backhoe, and stone dropped just inside the walls is conveniently placed for the machine operation.

Adequate access for concrete trucks is important when the slab is poured. Concrete trucks can enter the house through the garage, if it is level with the house, but additional access is needed. Ideally, there should be two openings for trucks in the south wall. At least one large opening will probably be left for floor to ceiling glass, allowing access to the living area end of the house. If the outer walls are of block, a temporary opening can also be left in the low wall at the bedroom end of the house. With poured walls concrete trucks must back up to the bedroom wall and extend their booms over it, into the building. Some wheelbarrow work will be needed, but only 10 to 15 percent of the slab area will be beyond the reach of the booms, and wheelbarrow runs into this area will be short, because concrete can be poured from any point along the south wall.

Once the slab has been poured insulation is applied to the inside of the structural walls, and the house is ready for framing. Framing goes rapidly, as lumber is dropped directly outside the low south wall and handed in, or brought onto the slab through openings left in the wall for the concrete trucks. The carpenters begin by working dry, on the slab, and the walls that form the inner shell and interior partitions can be marked out on the slab and built in a single operation. Temporary bracing is very simple, because the frame structure can be braced against the outer masonry walls until the roof is on. Time is also saved in constructing the roof, because the carpenters are working close to the ground. Ceiling rafters are set from low step ladders, and roof rafters and plywood sheathing are then stacked on the slab and handed up to crew members who are standing on the ceiling rafters.

After the house is framed the berms should be rough graded to simplify the remaining masonry work. This last stage in the construction of the building shell includes applying brick veneer to cover gable ends and eaves, completing the chimney and fireplace, and filling in access openings in the exterior walls. Interior walls used for passive solar storage should also be constructed at this point. Erected after the house is framed (so as not to interfere with the framing work) they must be built with block or brick that is unloaded outside the house. Most of the masonry storage walls will be located close to the south wall of the house, so cement blocks or bricks can be unloaded at any convenient point along this wall and handed in through its wide openings.

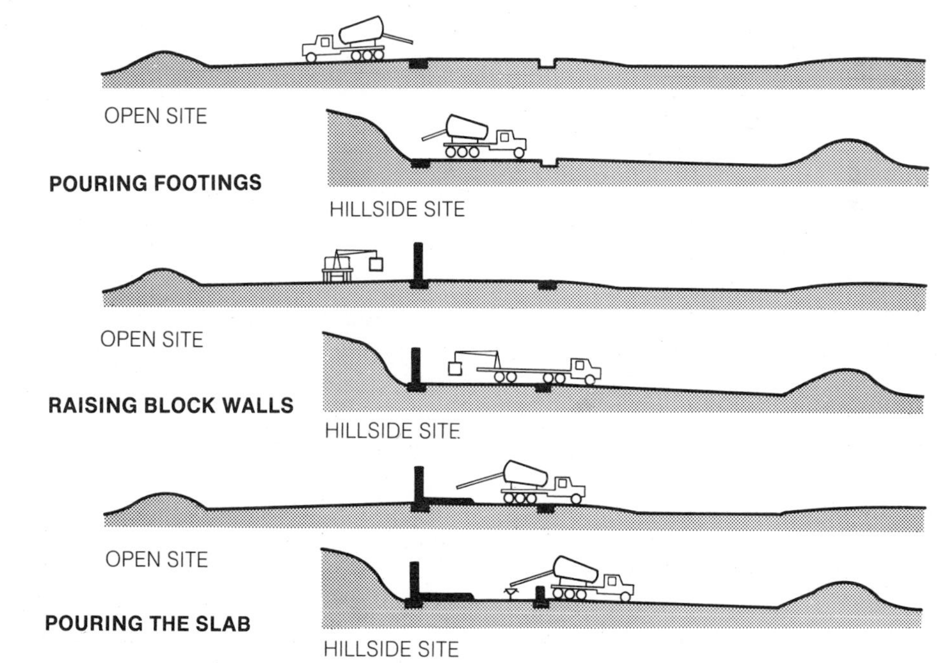

5–1. Building the walls and slab.

6

Wall and Roof Design

Designing to Resist Earth Pressure

The masonry walls that form the outer structural shell of the partially earth sheltered house will have a maximum of 8 feet of earth cover. This is only slightly more than the depth of a full basement, and it would appear that the earth sheltered walls could be built in the same manner as the ordinary basement, using 12 inches of block, or 8 to 12 inches of concrete. There are, however, basic differences between the two structures; basement walls do not resist earth pressure independently, but transmit loads to the opposite wall, creating balanced opposing forces from earth pressure with no overturning force on the walls. The horizontal or sideways thrust created by earth pressure is transmitted to the opposite walls by the floors of the frame house, which become a rigid diaphragm once the plywood or tongue and groove boards have been nailed down over the joists (figure 6–1A). It is commonly supposed that significant additional resistance to earth pressure is provided by the weight of the upper floors of the house. The weight of the upper floors does create a vertical load that opposes the action of the earth pressure, but the effect is small in comparison with the resistance provided by the diaphragm action of the floors.

The walls of the partially earth sheltered house resemble a basement structure with one open side. The floor of the ordinary open basement behaves like a truss, transmitting earth pressure from the back wall of the house to the end walls, which act as deep piers (figure 6–1B). The ordinary open basement forms a safe structure that can be as much as 35 feet long in light or moderately heavy soils, and 25 to 30 feet long in heavy, poorly drained soils. The unbraced length of the earth sheltered walls in the ESSPS house will frequently exceed these limits. In addition, thrust on the longer walls is resisted by the roof, and while the roof provides diaphragm or plate action, it will provide less resistance to earth pressure than the basement floor (figure 6–1C).

These structural conditions make it necessary to reinforce or buttress the walls of the ESSPS house whenever their unbraced length exceeds 16 to 20 feet. Piers will generally prove more economical than heavy reinforcing; unreinforced interior piers, 3 to 4 feet deep, provide effective resistance to earth pressure. If block is used for the walls, light horizontal reinforcing mesh should be laid between the courses to help transmit loads to the piers. The north side of the house contains the service spaces, and at most points along the north wall piers can be located at the required intervals without difficulty. Walls with full-height earth cover should also always be carried around the corner of the building for a distance of at least 3 feet; this will form a vital end pier for the wall. The interior piers will occasionally represent an unacceptable intrusion into the interior space and must then be located on the outside of the building. Exterior piers of either block or concrete should be tied into both a spread footing and the house wall. Block piers will also require vertical reinforcing.

Distances between piers can be increased by using light vertical reinforcement (figure 6–1D) made up of short, 10-inch bars set into hollow block that is filled with cement. The bars are pressed into place once the block has been filled, without being joined with wire wrapped around their ends, and they will not interfere with placement of the next course of block. The bars do not provide continuous vertical reinforcement, but

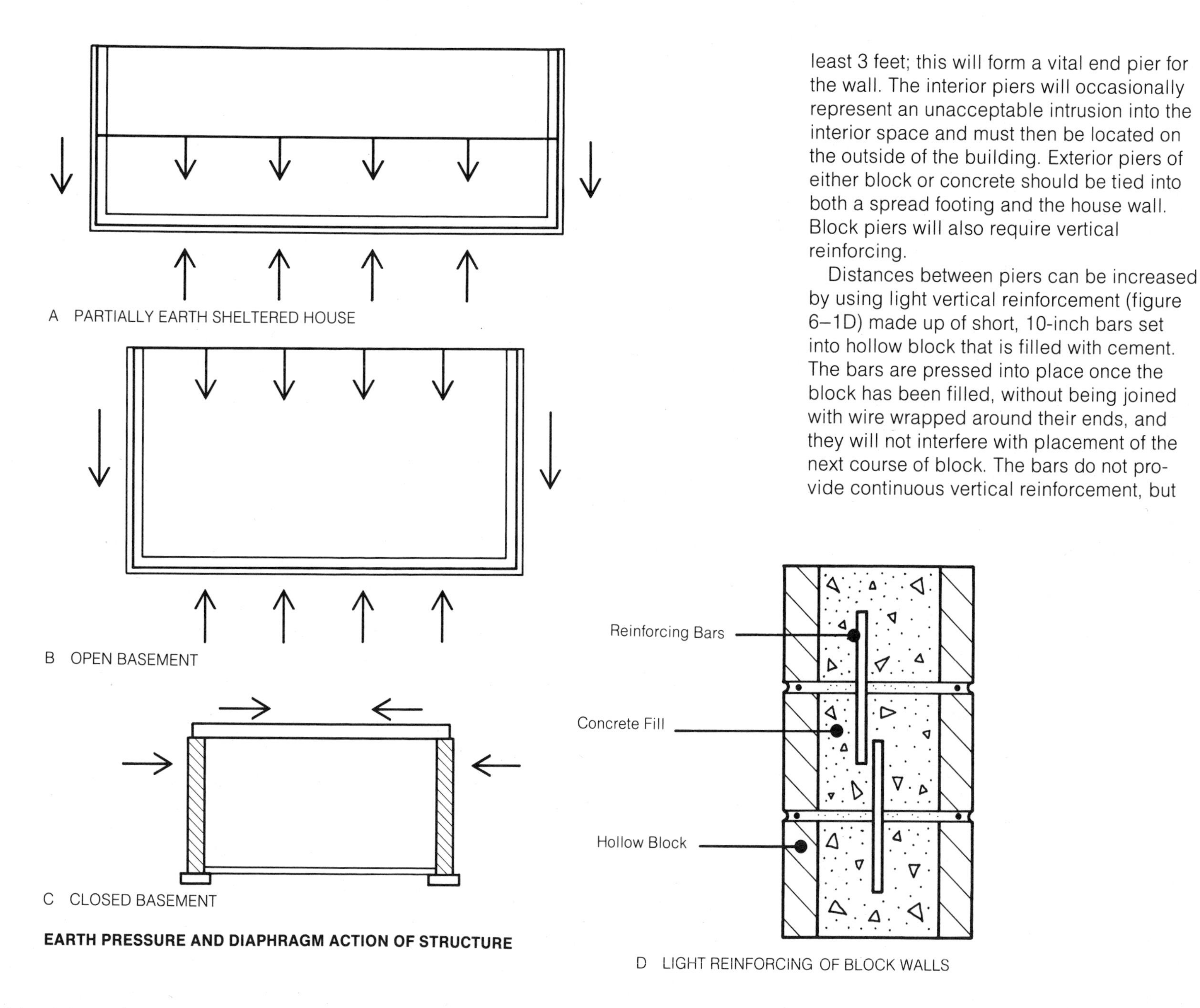

6–1. Resistance to earth pressure on masonry walls.

rather act, together with the cement that fills the cells of the hollow blocks, to increase the strength of the horizontal joints of the wall.

Earth pressure varies greatly in different types of soil. Heavy clay, which does not drain readily, will exert strong pressure against the walls once it becomes waterlogged. Sandy soils and earth that contains large amounts of shale drain more easily, and impose much smaller overturning loads on the walls. Because earth sheltered construction is relatively new, conventional building practices have not yet been established, and architects and builders should seek engineering assistance in designing the wall structure of the ESSPS house.

Waterproofing the Walls

The walls of the earth sheltered house must be waterproofed more thoroughly than the basement walls of a conventional house—water in the basement is a problem, but water in living areas is a disaster. A good waterproofing system has three components: proper grading, adequate drainage, and the waterproofing of the wall itself. All three elements are of equal importance, and it is a mistake to rely exclusively on the waterproof covering of the walls to obtain watertight construction.

Grading around the house must carry water discharged from the roof clear of the building and divert downhill runoff around the house. Earth berms will provide good drainage for the ESSPS house, but where the house is set into a natural grade a deep swale (V-shaped trench), not less than a foot deep, should be cut into the slope on the uphill side. The swale will carry runoff water around the house.

Careful grading, designed to keep water away from the building, should be supplemented by a stone drain at the base of the earth sheltered walls, as well as by stone placed under the slab. The exterior drain will cost several hundred dollars, and the layer of stone under the slab is also expensive, but both the drain and stone sub-base are important for adequate water protection.

With proper grading and drainage, the walls of the ESSPS house can be protected with a light-duty waterproofing membrane. Polyethylene is the obvious choice for the membrane material, since it is fully water resistant, very inexpensive, and requires few joints. Polyethylene is, however, easily torn, and its integrity can be destroyed by hidden damage during backfilling. This problem can be solved by using permeable polyethylene sold under the trade name Tyvec.

Tyvec was designed to provide a simple means of partially windproofing the conventional house. The microscopic holes that are formed as the material is built up in layers of very thin filaments allow water vapor to pass outward while partially blocking air infiltration, and it can therefore be applied to the outside of the walls of the frame house. Tyvec is also useful for below-grade waterproofing, since it has a great deal more resistance to tearing than ordinary polyethylene. Its microscopic holes, however, make it only partially effective as dampproofing; a two-layer membrane is therefore advisable. Moisture resistant black polyethylene forms the inner layer of the waterproofing membrane and is covered by an outer, protective layer of Tyvec. The cost of the material for the two layers at 1985 prices is only $0.08 per square foot.

The polyethylene sheets are installed by folding them into a masonry joint near the top of the wall, or by laying them over the top of concrete walls. (If the sheets are attached by nailing or cementing them to the face of the walls, they will pull loose when the wall is backfilled.) Corners should be lapped at least 12 inches and sealed with caulking beads. The bottom of the wall, where the joint connecting a block wall and its footing is located, and where hydrostatic pressure is at a maximum, is the critical point for any waterproofing system. In order to ensure a tight seal at the bottom of the wall, the joint between the walls and footing should be covered with smoothly formed cement, rather than left with the customary accumulation of cement drippings from the parging. Extra protection can be provided at the bottom of the wall by applying a strip of roofing cement that extends 6 to 12 inches up the wall from the top of the footing and 3 or 4 inches down the side of the footing (figure 6–2). The membrane is bedded in the roofing cement, providing a watertight seal at the bottom of the wall, as well as a resilient backing that will give extra protection to the membrane when stone for the drain is installed. Labor costs for the simple installation are low, and the waterproof membrane is very inexpensive.

Retaining Walls

Retaining walls join the partially earth sheltered house to the surrounding earth and will help determine the character of the entire design of the building. Retaining walls are costly to construct, however, since they must be built strong and must reach below the frost line. Stone is the ideal material for retaining walls, although the most practical choices will usually be reinforced concrete, cement block, and block faced with brick.

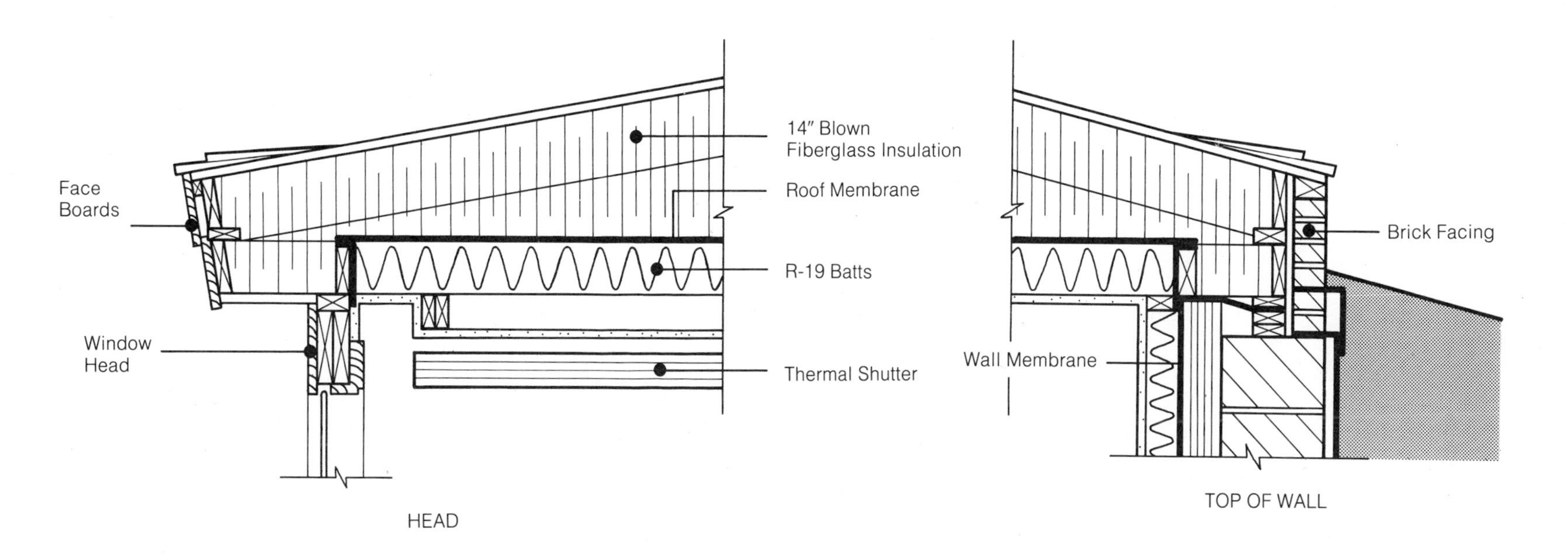

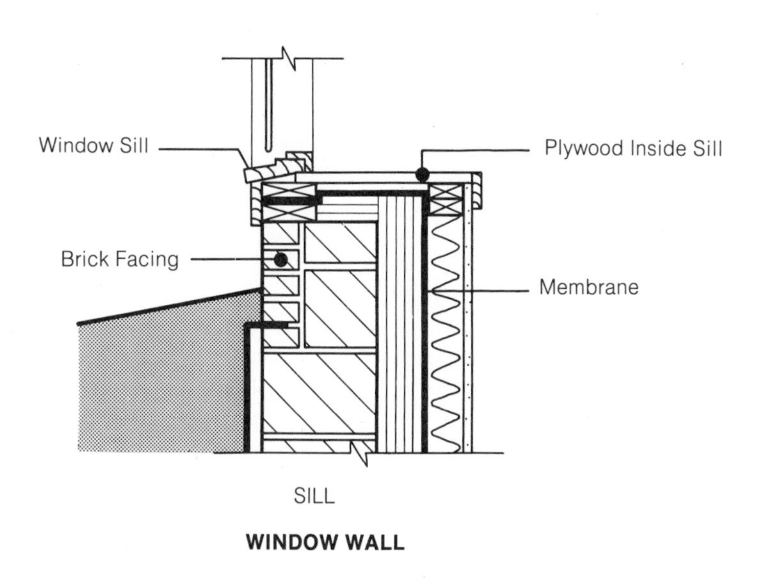

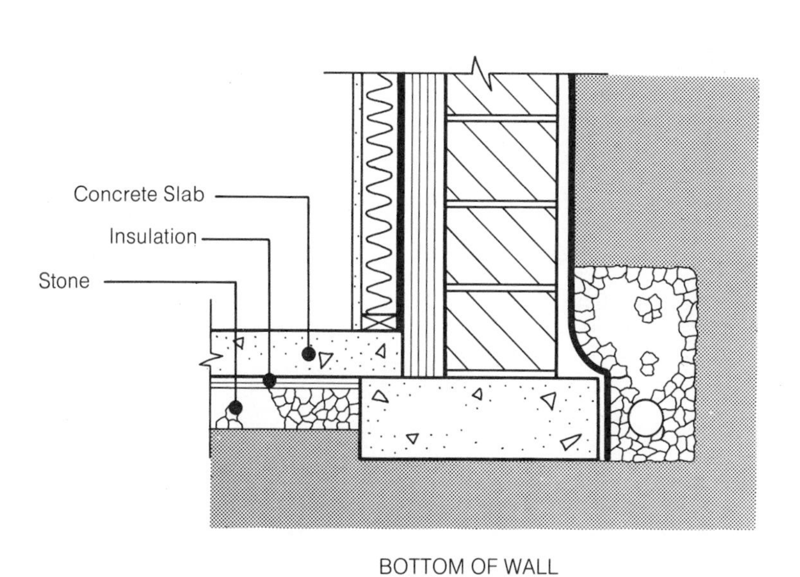

6–2. Construction details of the ESSPS house.

Reinforced concrete walls are extremely strong but are relatively expensive, especially when the house walls are block and the poured retaining walls are built under a small separate contract. Staining of exposed concrete walls is also a problem in northern climates. Rough-surfaced or "architectural" concrete resists staining, but its cost will almost always make this type of treatment impractical for small, poured-in-place walls.

Block walls, finished in stucco, are less expensive. Retaining walls of block must be reinforced or braced with piers. Hollow block should be used with cement fill and horizontal reinforcing set between courses. Reinforcing piers, including a return at the end of the wall, will be needed at intervals of 8 to 12 feet, unless the wall is low. The block retaining wall can be substantially upgraded by adding a brick facing. The facing increases the thickness of the wall from 12 to 16 inches, and the wall becomes much stronger. The brick finish will also be more durable and less likely to stain than either bare concrete or stucco.

Because the retaining walls are a major architectural feature of the partially earth sheltered house, the construction budget, even where modest, should include provision for well-built and attractive walls. The most effective method of controlling their cost is to use them sparingly. Wherever possible, earth should be graded down at the corners of the house and around the entrance, allowing walls to be replaced by landscaped banks (see figure 3–4). This will minimize both the length of the high walls, with their requirements for heavy reinforcement, and the exposed face area, so better finish materials can become part of the design.

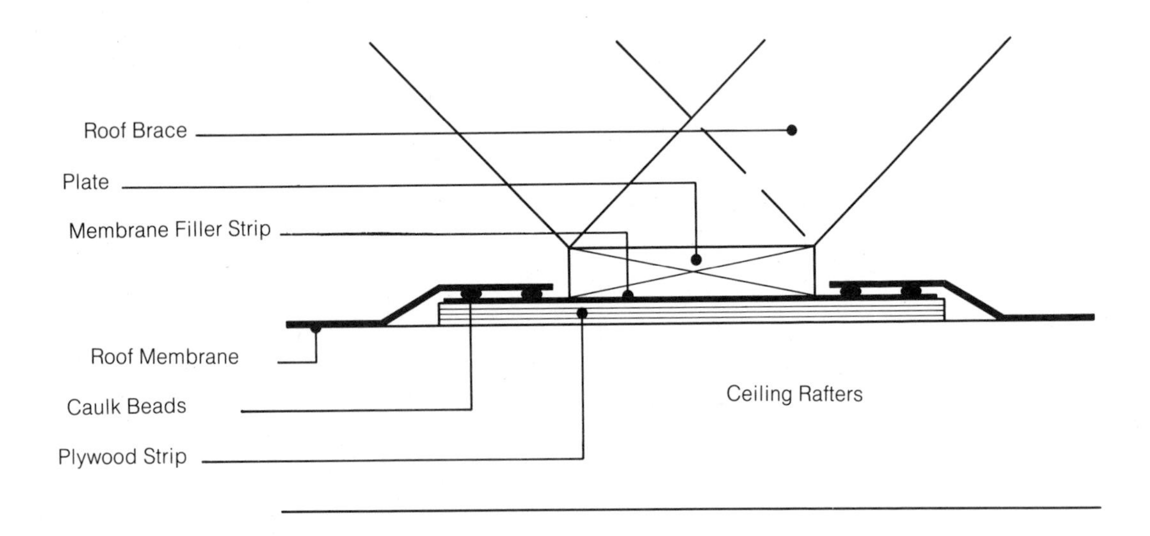

6–3. Detail of roof membrane joint.

The A-Roof

The standard low-slope A-roof requires little alteration when used as part of the ESSPS house, although it must be built with thicker eaves to allow adequate insulation at the wall line. The membrane and insulation installation present few problems provided membrane joints are properly detailed and care is taken to avoid damage to the membrane when it is being installed and insulated.

The membrane should not be installed until the roof framing is complete; this will require working in the low attic space as the sheets are spread out to the eaves after they have been rolled out across the rafters. A 40- by-100-foot roll of polyethylene is large enough to cover most houses, but in many designs the membrane will be interrupted by changes in level and roof braces. Sometimes piecing the membrane will also be required by the geometry of the plan. Roof joints, in contrast to those in the wall, are exposed to the wind and must be fully airtight. In addition to careful joint detailing it will be useful to back up the membrane with Tyvec, which can be used in place of the roofing paper normally laid under asphalt shingles, providing partial protection against air infiltration and supplementing the inner membrane. The material is not recommended by the manufacturer for roofs, because it is slippery, but it can be used on low-pitched one-story roofs with little risk.

Roof membrane joints should be formed with a lap joint that has a narrow strip of plywood or 1 × 6-inch roofers as a base. The ends of the overlapping sheets are joined with one or more caulking beads, and the sheets are pressed together against the base strip for a secure joint. Heat sealing can also be used if the equipment is available. Where rafter braces interrupt the membrane, a strip of plywood is laid across the ceiling rafters and covered with 12-inch-wide polyethylene before the braces are nailed in place (figure 6–3). Once framing is completed and the membrane is installed, the separate sheets on either side of the braces are sealed to the strip of polyethylene that was put in place earlier, with the wood strip again forming the base for a secure joint.

The joint with the wall membrane at the gable ends of the house is very simple. Joining the roof and wall membranes on the long sides of the house is more complicated, since here the rafters are in the way. This joint can be formed by placing connecting strips of polyethylene between the rafters. The strips are backed by blocking and are held in place with two small patches of roofing cement. Caulking can then be applied to make a tight seal between the rafters and the membrane (figure 6–4).

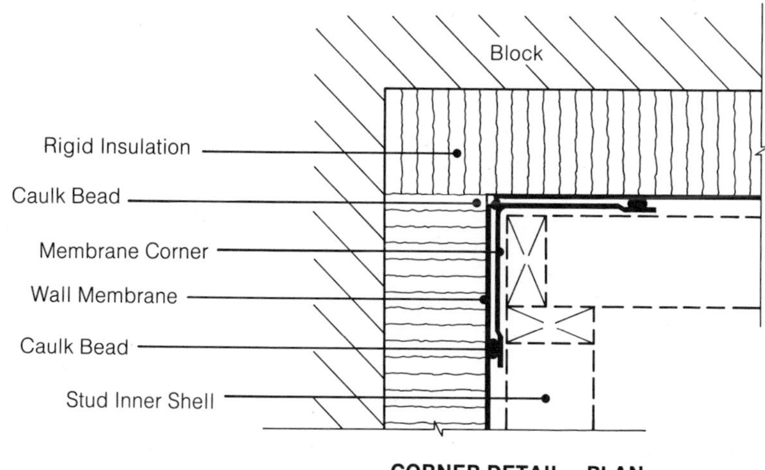

CORNER DETAIL—PLAN

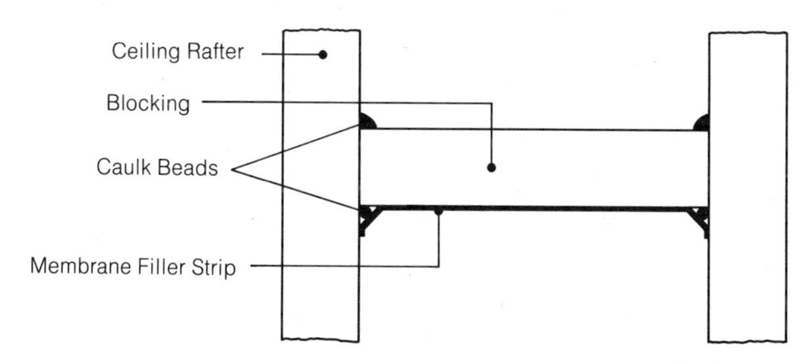

PLAN OF MEMBRANE JOINT AT BEARING WALLS

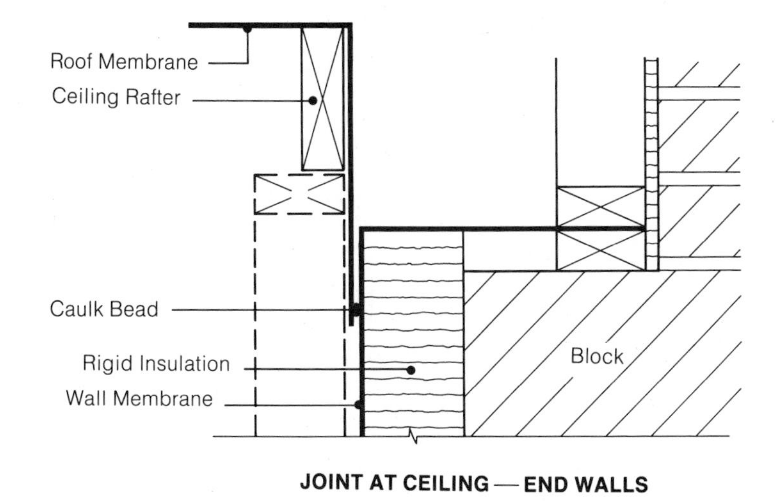

JOINT AT CEILING — END WALLS

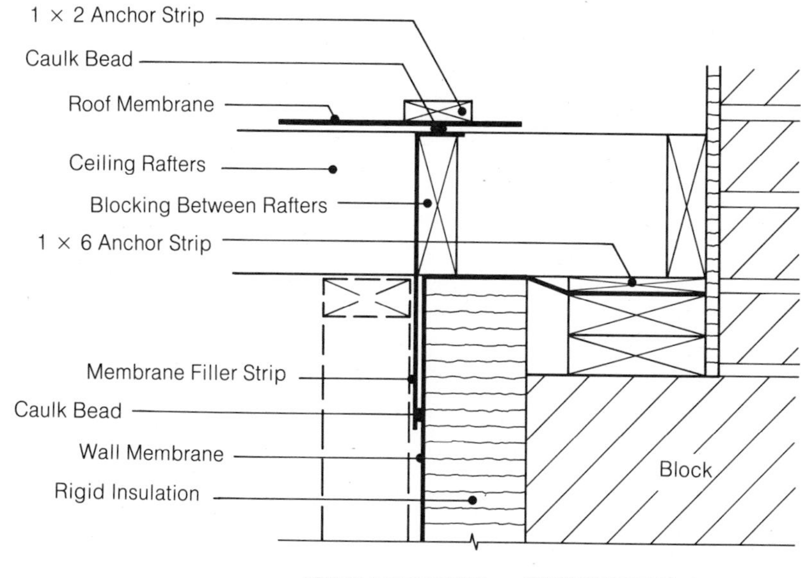

JOINT AT CEILING— BEARING WALLS

6–4. Wall and roof membrane details.

Blown insulation is installed after the membrane is in place. The installers work down the middle of the roof, blowing the insulation to either side. Access for this operation is from the gable ends. Fiberglass batts, installed between the roof rafters, complete the roof insulation. Breaks in the batt insulation from wiring, electrical fixtures, and plumbing lines are not important, since they are covered by the heavy upper layer of blown insulation. Interruption of the insulation blanket by branch heating and ventilating ducts is more significant, and bays that carry duct runs should be covered by 1 or 2 inches of rigid insulation laid on top of the membrane before the blown insulation is installed.

Skylights

Skylight openings should be kept as simple as possible to avoid complicating the roof framing. The easiest way to install the skylights is to set them parallel to the roof, on low curbs. In this position they are nearly horizontal and their primary function is to admit light into the interior of the house, rather than to provide an additional source of passive gain. Insulated knee walls are required around the skylight openings, and the vapor barrier must be carried up to the skylight curb (figure 6–5).

Winter passive gain from skylights will be nearly tripled by constructing dormers, which allow the skylights or glass to be set at an angle of 55 to 60 degrees from the horizontal. In regions that receive less than normal amounts of winter sunlight, large dormers will often be the most economical method of increasing passive gain.

With a 2/12 roof slope, light wells in the center of the building are typically over 4

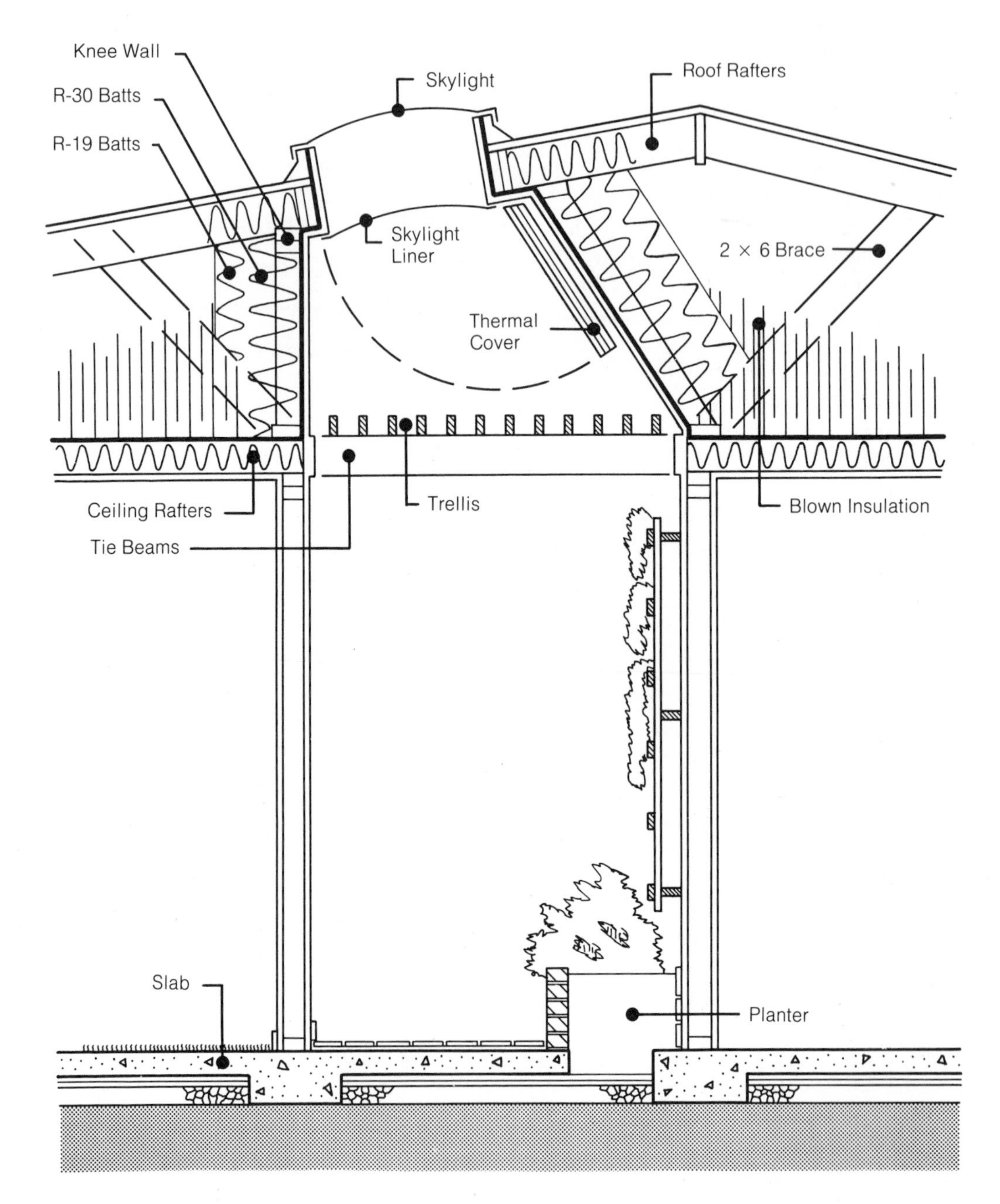

6–5. Section through the gallery.

feet deep. Abrupt changes in ceiling height are not usually desirable in the ESSPS house, since its spaces have flat ceilings, and generally there are limited changes in room height. This will tend to make a sudden vertical expansion of the space under the skylights an intrusive design feature, which can be avoided by building openwork ceilings that allow sunlight into the central corridor while cutting off the view up into the light wells.

In a central corridor designed as an indoor garden, the open ceiling can take the form of a trellis. Lengthwise members, lapped at their ends, are placed on top of heavier cross beams, which serve as ties for the ceiling rafters and can also be used to box in electrical lines and branch-heating ducts. The carpentry work is simple; lumber and wood finishing make up most of the cost of the trellis. A trellis covering an area 30 to 35 feet long and 5 or 6 feet wide costs only $500 or $600 dollars at 1985 prices and will be a major decorative feature of the design. If necessary, cost can be reduced further by building separate trellis sections under the individual light wells, in place of the continuous ceiling.

7

Windows and Window Protection

Window Design

The practical requirements of the ESSPS house greatly influence the design of its windows. These must be nearly airtight, have frames with good thermal resistance, must be sized to provide adequate winter passive gain, and most of the window area must face south. The windows must also be designed for economical installation of thermal shutters.

The south-facing wall will be over one-half glass. Aluminum sliding windows and patio doors are the least expensive ways to incorporate large glass areas into the house. They will not, however, be satisfactory when used in the energy-efficient building, because even the best aluminum units have relatively high rates of thermal transmission and air infiltration. Wood frames must be used throughout, for operating sash, fixed lights, and doors that provide access to the garden.

The number of operating sash should be kept to a minimum and most of the glass area made up of fixed lights. This will produce a tighter wall than if a large number of operating sash are used. Reducing the number of operating sash will also reduce construction costs, since good quality wood sash are more expensive than fixed glass set in wood frames. Fixed glass should be installed in large lights. Large lights require heavier, more expensive glass, but this is offset by savings on the cost of trimming the openings. The labor needed for trimming an opening is virtually the same whether the window is large or small, so large lights will save on labor costs. In addition, less trim material is required if the number of openings is reduced.

The large fixed lights are combined with casement or awning windows. These windows are preferable to double-hung windows, because their entire area can be used for ventilation. If the house is air-conditioned, one large window in each room will provide adequate ventilation for spring, fall, and cool summer weather. If, however, moving air drawn into the house by fans is substituted for an air-conditioning system, additional operating sash will be needed. The cost of the extra operating windows should be considered when comparing the cost-effectiveness of air-conditioning and ventilating fans.

Window openings can be designed as horizontal or vertical units (figure 7–2). Glass is less expensive for vertical fixed lights, because inexpensive mass-produced

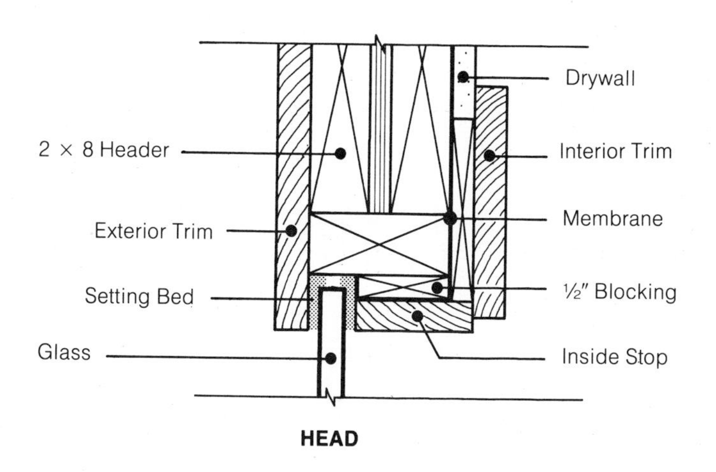

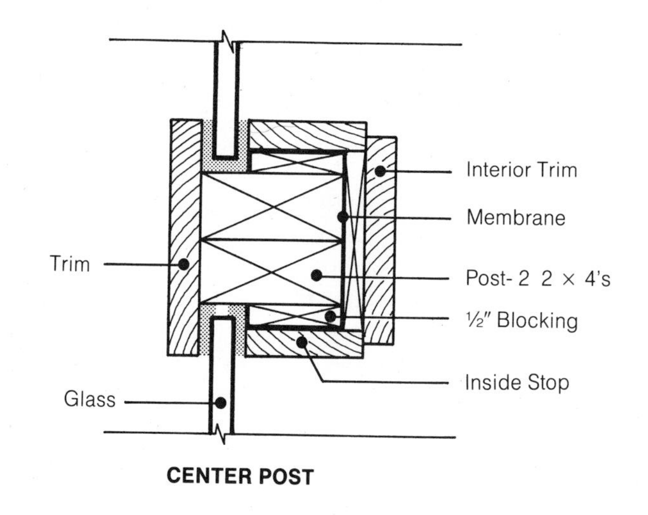

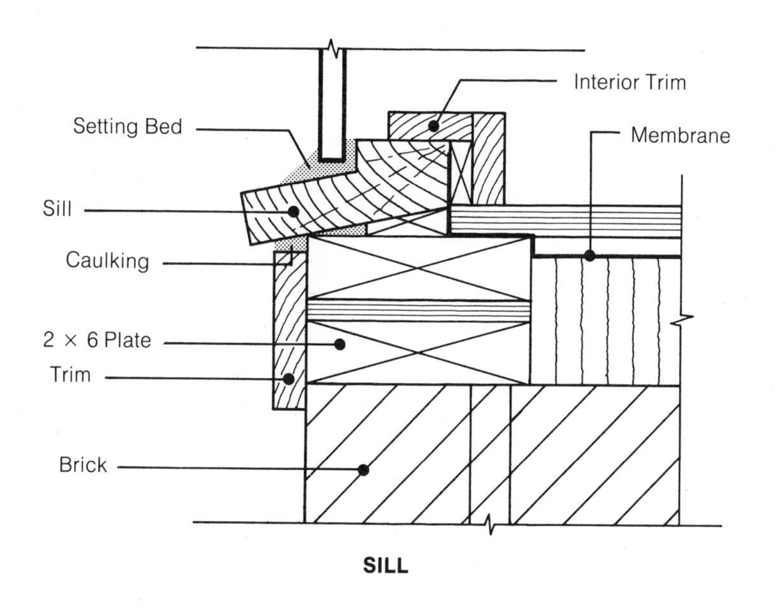

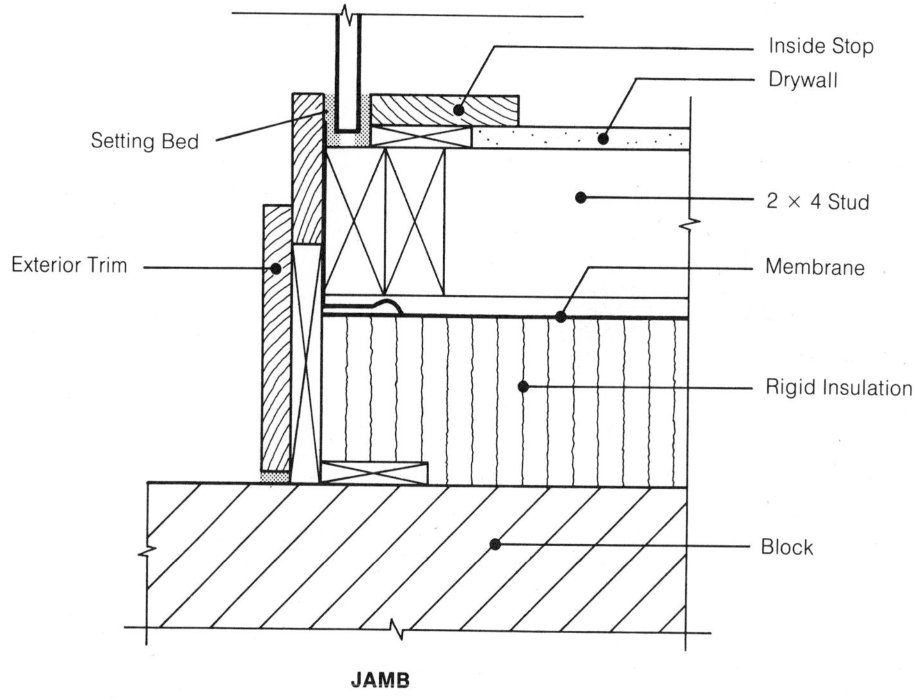

7–1. Window details.

patio-door glass can be used. The overall cost of the wall, however, will be lower with horizontal openings. Berms protect the wall below sill height with a horizontal opening, eliminating the need for a deep foundation. In addition, very large top-hinged shutters can be used to cover horizontal openings, while vertical windows require more cumbersome and expensive bifold or sliding shutters. When economy is the primary consideration, most of the glass should be placed in a horizontal band that runs continuously or nearly continuously across the south side of the house. This band can be extended as needed around the east and west sides of the building. Vertical openings should include doors opening out into the garden, and should be restricted to one or two rooms. Given a more generous budget, the window wall can include more extensive use of vertical openings, allowing the designer to emphasize the relationship between the earth sheltered house and the surrounding landscape.

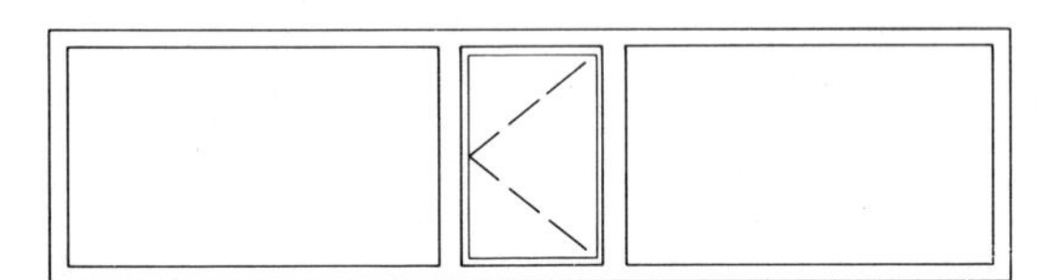

HORIZONTAL GLASS

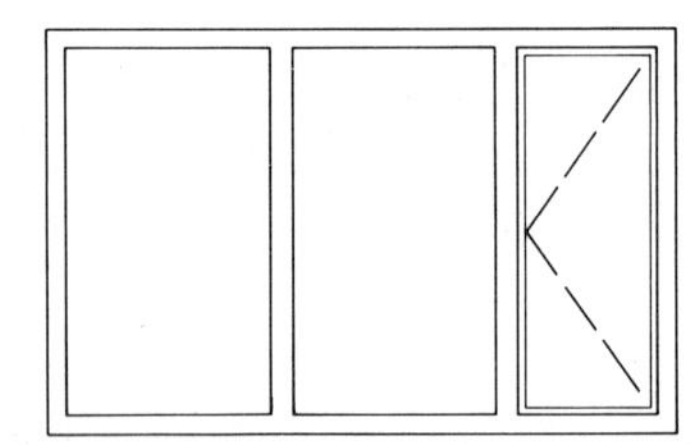

VERTICAL GLASS

7–2. Window elevations.

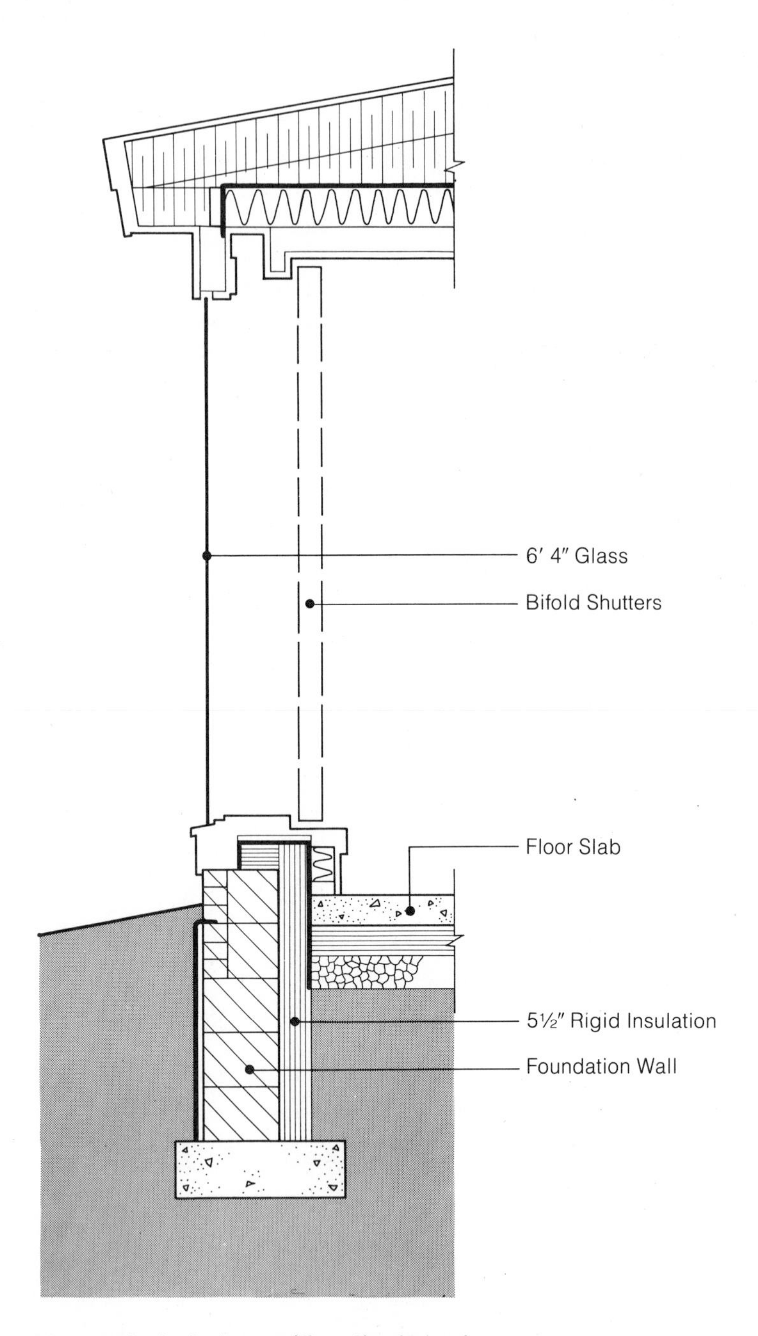

7–3. Wall section showing vertical windows with patio-door glass.

Thermal Shutters

The ESSPS house is built with R-34 or heavier walls and its roof insulation has a minimum rating of R-57. In order to complete the heavy thermal envelope, windows must be protected at night by coverings that provide thermal resistance in the range of R-15 to R-20. Economical factory-built shutters that provide protection of this type are as yet only available in a limited range of sizes, and most of the shutters must be field fabricated.

Lightweight thermal shutters can be built with rigid insulation set into a wood frame. If 3½ inches of polystyrene are used, the shutters will have a rating of R-17, and total protection provided by the glass, airspace between windows and shutters, and the shutters themselves, will exceed R-20. One side of the shutter should be covered by ¼-inch or 3⁄16-inch plywood for rigidity; the other side can be spray painted or covered with fabric. The shutters must also be designed to form an airtight seal around the window when they are closed, to prevent condensation from warm room air filtering into the space between the shutters and the glass.

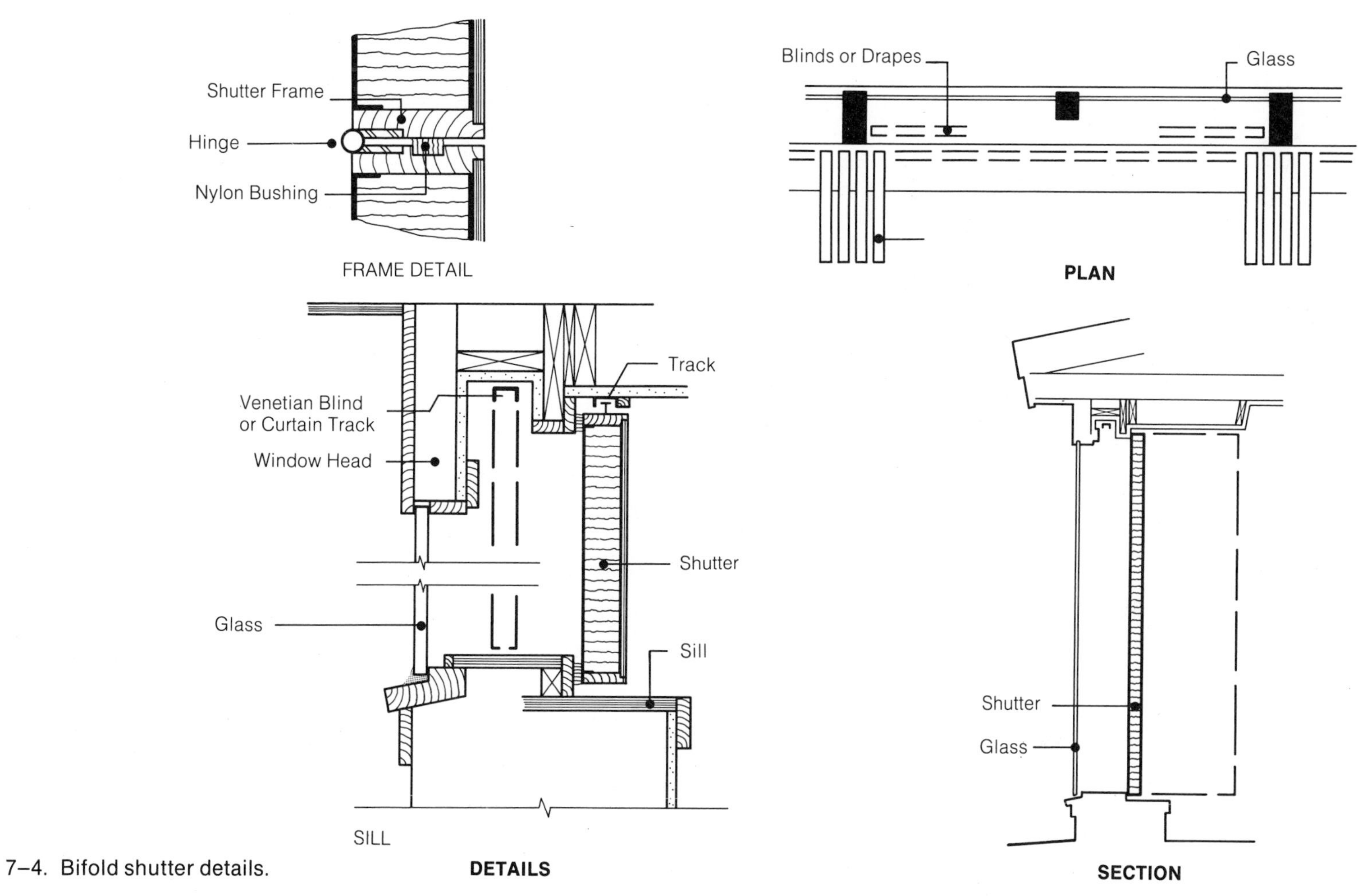

7–4. Bifold shutter details.

Small window openings can be covered with pairs of sliding or hinged shutters, although these shutters take up 50 percent of the wall when opened. Wide continuous openings require bifold, sliding, or top-hinged units (figure 7–4). The top-hinged unit is the most economical, because it can be built in almost any length, and a single shutter will cover the entire outside wall of all but the biggest rooms of the house. Most of the cost of the shutters goes for the labor of fabrication and installation, so large units are less expensive than smaller ones.

The overhead shutter is operated with the aid of a pivoted counterweight placed in the wall at one end of the window opening. Correctly balanced, the shutter presses lightly against its rubber or nylon seals when it is in the down position. In the up position, it is held in place by nylon line run through the ceiling rafters to the counterweight mechanism. A dropped ceiling provides a neat housing when the shutter is in the raised position and also creates a recess for venetian blinds or a drapery track.

Placement of the counterweight mechanism will sometimes be a problem and, in addition, the overhead shutter cannot be used where vertical lights extend the glass area close to the floor. When the overhead installation is not appropriate, sliding or bifold units can be used. Care must be taken in detailing and installing the multipanel units in order to provide tight closure. Stacking open shutters will also be a problem, particularly with bifold units. R-17 polystyrene shutters are over 3 inches thick. When two adjacent four-panel units are opened, a stack of four units, over 15 inches wide, is formed. Pilasters can be designed to house the open shutters, but in many instances large pilasters will be intrusive when placed along a window wall. This problem can be solved by using a more expensive insulation material such as Thermax in place of polystyrene or Styrofoam. One and one quarter-inch Thermax has a rating of R-9, so total protection with the shutters closed is an acceptable R-12, while integration of the bifold shutters into the window wall is easier than if bulkier materials are used.

Sunscreen Details

A protective system that combines full summer shielding with unobstructed glass area during the cold months can be designed for the south wall. A 12-inch roof overhang provides weather protection for the windows without blocking sunlight during late fall, winter, and early spring. The overhang can be extended another 3 feet with removable sunscreens, creating a 4-foot-wide sunshield that will provide almost complete protection from direct summer sun in medium latitudes (figure 7–5).

Removable wood or aluminum sunscreens can be built in lengths of 6 to 8 feet. One person installs the screens from the ground, sliding them into brackets attached to the barge board and window header (figure 7–6). The brackets are detailed so that a tight fit is not necessary, and the lightweight screens will slide smoothly into place. Sunscreens can also be set into the openings of a trellis formed by extending the roof framing, which makes their installation still simpler, although the permanent trellis will block some sunlight during the winter months. The chore of putting up and removing sunscreens at the beginning and end of the summer can also be simplified by storing the screens on a wheeled cart; a path built through the garden will allow the cart to be drawn around the house. Most of the windows have berms coming up close to their sills and the sunscreens are installed at shoulder height.

East- and west-facing glass is more difficult to protect from direct radiation than south-facing glass, because overhead screens do not block low-angle early morning and late-afternoon sun. Light-colored interior blinds will reflect over 50 percent of the direct radiation falling on the glass. Deciduous trees planted 30 to 60 feet from the house will provide additional protection. Their shadows will reduce winter passive gain, but in most climates this will be more than offset by the smaller summer cooling load.

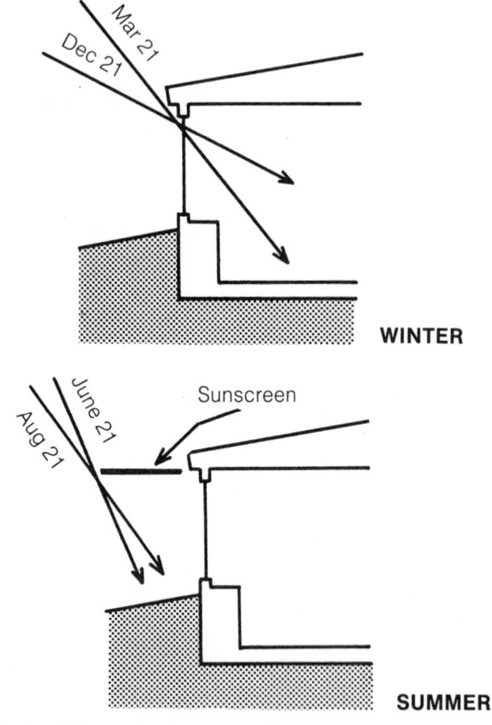

7–5. Winter and summer sun angles on a south-facing surface.

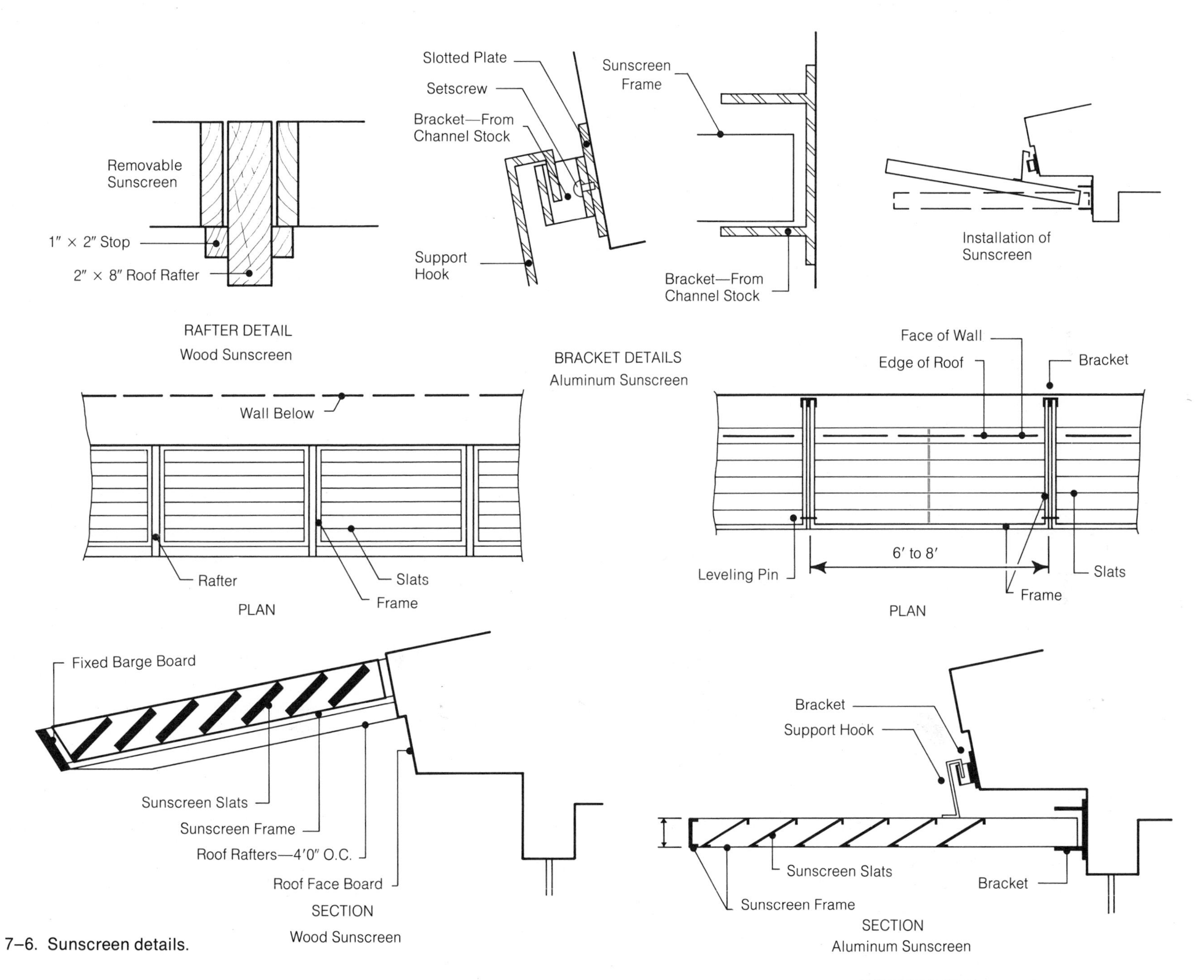

7–6. Sunscreen details.

8

Passive Solar Storage Methods and HVAC Systems

Passive Storage and Hot-water Preheat

The medium-size ESSPS house typically requires a little over 160,000 Btu of passive solar storage capacity (see Part III). A partially carpeted slab, together with walls and ceiling, provides half this capacity; only slightly more than 80,000 Btu of additional storage capacity must be installed to complete the system.

Nonstructural masonry walls are a proven method of providing storage capacity. Most of these walls should be located in the carpeted areas of the house in order to compensate for reduced slab storage here and distribute capacity as evenly as possible. The dividing walls between the bedrooms are a logical place for the storage walls, which are most efficient when they are 8 inches thick and exposed on both sides. Storage capacity should also be provided in a living room with a carpeted slab. An interior fireplace can be expanded into a storage mass, or if the fireplace is on an exterior wall, the whole wall can be covered with 4-inch brick veneer to form a decorative storage element.

Storage capacity can also be provided by phase change materials (Glauber's salts), which absorb large amounts of heat when changing state and can be compounded so that the change of state occurs within the normal temperature range of the ESSPS house. One currently marketed phase change system packs Glauber's salt in 22-inch-long stainless steel tubes, each of which will absorb or discharge 2,000 Btu when changing state. Using the tubes to their full capacity requires forced air flow; in the ESSPS house the tubes are stacked in a plenum through which air is moved by the furnace motor of the backup heating system.

Phase change materials provide storage at a lower initial cost than masonry walls, but have a limited useful life. Until recently, the deterioration in the performance of Glauber's salt was very rapid and its useful life short. Improved versions of the material, for which a useful life of twenty years is claimed, are now available, but manufacturer's claims have not yet been verified by long-term field testing. With a twenty-year useful life, the cost of amortizing the phase change materials is only 60 percent of the cost of amortization of cement block walls with equivalent storage capacity. Even with a ten-year useful

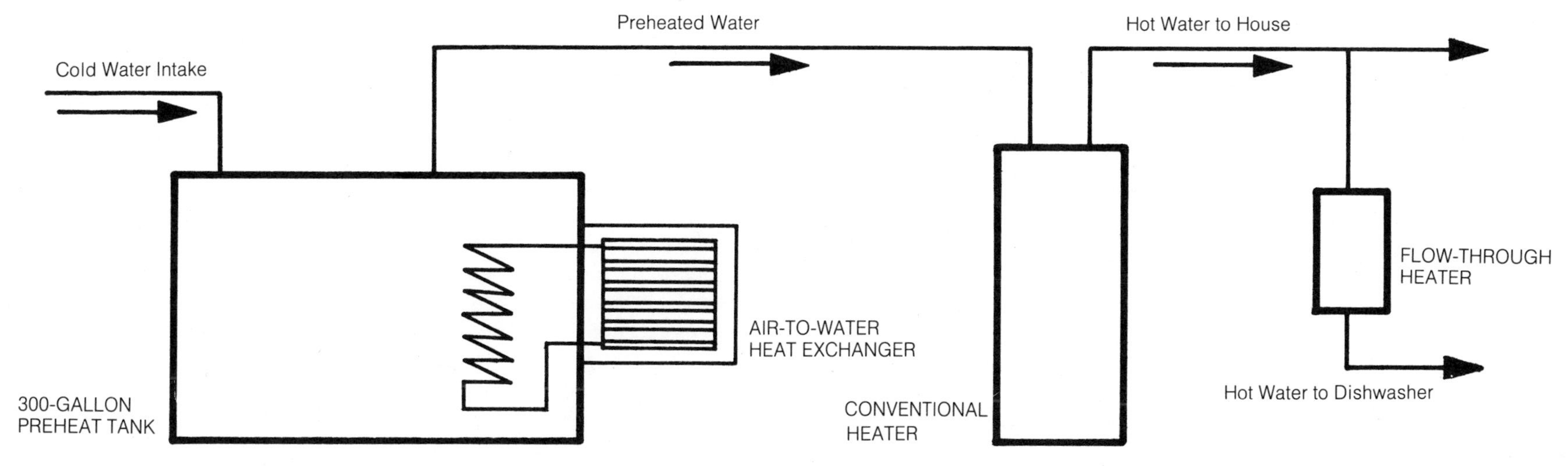

8–1. Schematic layout of hot-water supply system.

life, the amortization cost of the phase change materials will be somewhat lower than that of the block walls. Because the durability of the system is not yet fully proven, conservative design will not place excessive reliance on phase change materials and will instead combine use of masonry walls with the newer and potentially more economical storage method.

On clear days in winter, and during clear and partially clear days at other times of the year, passive solar input will exceed the storage capacity of the system, leaving surplus energy available for hot-water preheat. The preheat system includes a large storage tank, capable of providing a four-day supply of preheated water, an air-to-water heat exchanger, and a small air-handling unit, which draws air from the living spaces through the heat exchanger. The exchanger is attached to the preheat tank. Tank water circulates through the exchanger and is warmed by the room air (figure 8–1). In winter, room temperature is allowed to reach 75° F before excess passive gain is dumped, and water can be preheated to 73° F in the exchanger. During the summer, room temperatures will normally be higher, even with central air-conditioning. The energy-conscious family probably will not set the thermostat below 80° F in warm weather, so water can be preheated to 78° F in the summer months.

After preheating, the water is raised to its discharge temperature in a second heater, which can use gas, electricity, or solar panels as its energy source. Maximum savings are obtained with preheating if the final discharge temperature is kept low; the hot-water system should be designed to discharge water at approximately 110° F instead of the usual 135° to 140° F. The energy saved from lowering the discharge temperature will be reduced somewhat, because water supplied to the dishwasher must be heated to 135° F, using a small direct-flow heater; water at 110° F is hot enough for all other uses.

Tank losses from preheated water stored for several days during winter, when surplus passive energy is available only intermittently, are very small because the preheated water is at almost the same temperature as ambient room air. In winter, when inside temperature is allowed to drop to 67° F as the passive storage system is depleted, the maximum temperature differential between the tank water and the room air will be only 6° F. If 6 inches of batt insulation are wrapped around the tank, storage losses will be negligible.

The entire hot-water system, including the preheat tank, air-handling unit and heat exchanger, main secondary heater, and direct-flow heater, is relatively expensive in comparison with most of the other equipment in the ESSPS house. It is cost-effective, however, because it uses surplus energy during the entire year, and in summer it both reduces hot-water heating costs and the load on the cooling system.

The Heat Exchanger

The sealed dwelling requires 0.4 to 0.5 air changes per hour for comfort and habitability. The ESSPS house has an extremely low rate of air infiltration, and most of its ventilating air is drawn into the building through a heat exchanger. This allows the outside air to be warmed by exhaust air that is directed through the exchanger. The air-to-air exchanger recovers approximately 80 percent of the heat of the exhaust air. It uses, however, a substantial amount of electricity to operate the fans that force the intake and exhaust air through the narrow confines of its baffles. Lowering the amount of electricity that the exchanger consumes would make the ESSPS house still more efficient; exchanger manufacturers have, however, emphasized high heat transfer rates and low initial equipment cost and paid less attention to energy consumption during operation.

The energy consumed by the exchanger does not vary with the temperature differential between indoor and outdoor air, and the heat exchanger will only reduce heating costs when there is a sufficient difference between indoor and outdoor temperatures. Current generation counterflow exchangers save little energy when the outside air temperature is over 40° F, and above 45° to 50° F most units will consume more energy than they save by recovering heat. When outside air temperature is above the break-even point, ventilating air should continue to be drawn into the house through the exchanger system in the amount required for proper air circulation, but it should be routed around the exchanger baffles. In mild weather, when indoor and outdoor air temperatures are almost the same, fresh air can be admitted through the operating windows. If the house is air-conditioned, it is sealed in hot weather and ventilating air is drawn into the house through the exchanger system, again bypassing the heat-transfer baffles to minimize energy consumption.

Two different types of ventilating systems can be used with a heat exchanger. Some systems draw exhaust air only from the kitchen and bathrooms. Usually their exchanger operates when the kitchen hood and bathroom fans are turned on, although sometimes control is provided by a humidistat, which turns the exchanger on when indoor humidity reaches a preset level. These systems are best suited to small houses, under 1,600 square feet. They emphasize ventilation of the kitchen and bathrooms, because these spaces are the principal sources of odors as well as of much internally generated humidity. With either control, the operating time of the exchanger is minimized, relatively few fresh air outlets and return intakes are needed, and ductwork runs are very short.

More elaborate ventilating systems are designed to supply fresh air directly to every room in the house; systems of this type are usually operated at regular intervals whenever the windows are closed. In a medium-sized house, these systems require approximately 120 feet of ductwork. The exchanger system will normally handle only 200 cubic feet per minute (cfm) of air—one-tenth of the volume of air moved by a warm-air heating system in a conventional house of the same size—and its ducts are very small. Inexpensive 6-inch round pipe can be used for trunk ducts, and 3-inch pipe will be sufficient for the branch runs.

The volume of ventilating air is controlled with a timing device. Most counterflow units are designed to operate with maximum efficiency at about 200 cfm. A 2,000 square foot house will require slightly less than 100 cfm of ventilating air, and the exchanger will normally run approximately thirty minutes in each hour. The exchanger system must also be able to respond properly when the kitchen hood and bathroom fans are turned on, since these fans draw air from the house, and when the fans are on, the exchanger should operate continuously.

When the expense of operating the exchanger is added to the cost of amortizing the exchanger, current generation heat-exchangers are only marginally cost-effective in moderate climates. But with even a small increase in energy prices, the exchanger becomes cost-effective, and if high rates of increase in energy costs prevail, the exchanger will produce very substantial cost savings for the homeowner.

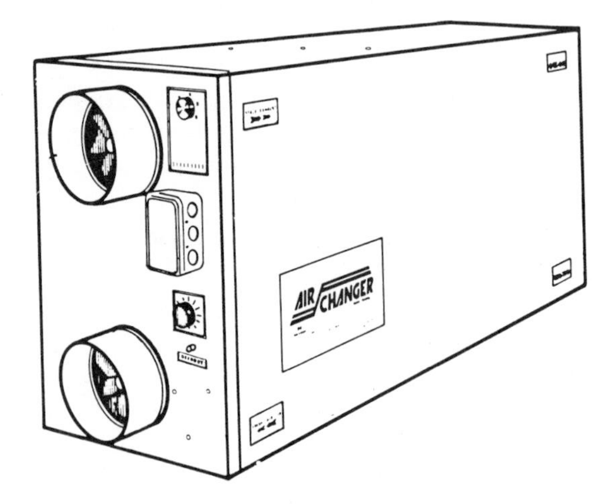

8–2. Counterflow heat exchanger. (Courtesy of The Air Changer Company, Ltd.)

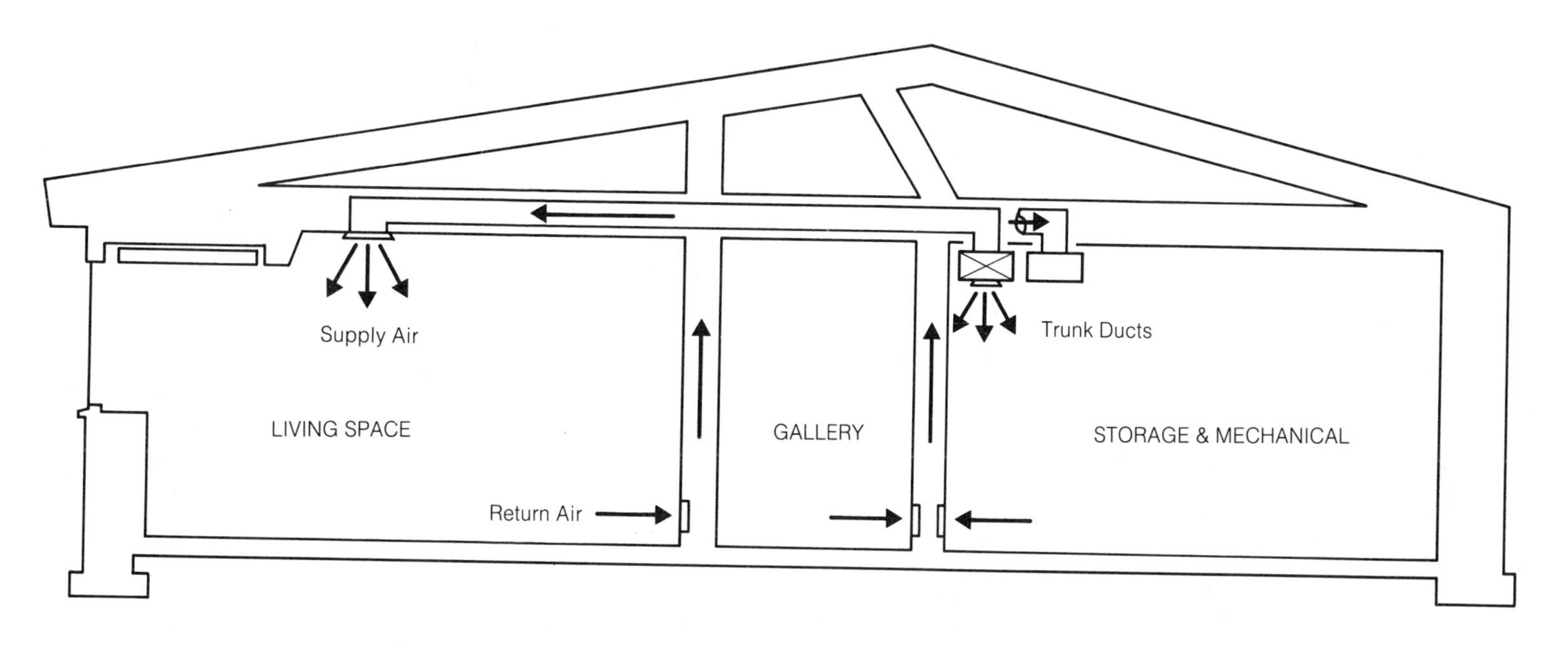

8–3. Air circulation and ductwork.

Controlling Internal Heat Sources

Internal heat sources contribute a significant amount of energy during the winter, and because this is well understood, devices that retain the warm exhaust air of the laundry dryer in the building space, instead of venting it, are now in general use. Less attention has been paid to discharging heat generated by internal sources during the summer months. The energy-efficient house should include exhaust systems for both the oven and refrigerator. These exhausts should, like the dryer exhaust, be valved so that they only discharge heat from the house in warm weather. A dishwasher exhaust will also be cost-effective if the machine is properly insulated. Controlling latent heat sources (water vapor) is more difficult although it is useful to install bath and shower doors that extend to the ceiling, with an exhaust fan in each enclosure.

Backup Heating and Room Air Circulation

The backup heating system that must be installed in the ESSPS house is not an emergency device. It is intended to operate at regular, if infrequent, intervals during the heating season. The air-handling unit of this system circulates air when the house is acquiring passive gain and when the storage system is discharging heat. The backup system will frequently be combined with central air-conditioning. A winter/summer heat pump for a 2,000-square-foot house will have a capacity of 1½ to 2 tons, compared with 3 to 3½ tons in a conventional house.

Air distribution with the smaller heat pump must be designed to ensure proper air movement in all parts of the house. The air distribution system should have at least one supply outlet in every living space. The supply ducts run to ceiling diffusers, which are placed toward the outside walls of the rooms, but which can be held back from the windows to avoid interference with the thermal shutters (figure 8–3).

Return registers, located close to the floor on the interior partitions, must be provided in every room as well as in the halls. This assures constant movement of air throughout the rooms, since the diffusers throw the supply air toward the outside (window) walls. Return air rises through the stud spaces of the partitions and enters short horizontal branch ducts at the ceiling. Both the return and supply branch ducts are connected to trunk ducts that run the length of the house; 4-inch and in some cases 3-inch pipe can be used for branch runs. The total length of the ductwork runs is relatively long, but small duct sizes, combined with a very simple installation layout, make the duct installation comparatively inexpensive.

Rapidly rising electricity costs in many parts of the United States have caused increasing attention to be focused on cooling systems that use less energy than the power-hungry air conditioner. Circulating fans are the best alternative for the ESSPS house. Ceiling fans, which have come back into use after vanishing nearly thirty years ago, save a great deal of energy and are inexpensive to install. Roof-mounted fans that draw air in through open windows will provide more rapid air movement than the circulating fans. Two fans, however, are usually required, and air must be vented through large louvers installed in towers that project above the roof. Roof fans are energy efficient compared to central air-conditioning, but their initial cost is greater and the cost-effectiveness of the fans will depend on local summer climate and electric rates.

PART III

PERFORMANCE ANALYSIS AND PASSIVE SOLAR DESIGN

9

Winter Heating Performance

Local Climatic Variation

The performance of the ESSPS house and the size of the fuel savings that it generates depend to a significant extent on local weather conditions. The optimum design will also vary, since weather conditions affect the thickness of insulation, size of windows and skylights, and passive solar storage requirements. The design of the ESSPS house must be evaluated on the basis of climatic factors, including winter and summer temperatures, the availability of sunlight, daytime and nighttime temperature variation, and the pattern of alternating clear and cloudy days.

These considerations make it necessary to study design and performance in a specific location, using local weather data. Methods for determining energy consumption and energy savings are analyzed here for a four-bedroom house with 2,000 square feet of living space, located in southeastern Pennsylvania (5,100 heating degree days and 500 cooling degree days). This analysis can readily be applied to other regions in the United States and Canada.

Space-heating Loads

Space-heating loads for the ESSPS house should be compared to a conventional house built recently (1985) rather than to a house that conforms to older standards. The conventional house used for comparison here has added wall insulation, good quality, double-glazed windows, and the recently developed Tyvec insulation barrier. These changes significantly reduce building heat loss compared with a house constructed a decade earlier.

The ESSPS house has a much lower hourly rate of heat loss than the upgraded conventional house. Heat loss is one-half as great during the day, and one-fifth as great at night, when its thermal shutters are closed.

Table 9–2 shows ESSPS heat loss with 0.4 air changes per hour. Ventilating air drawn through the exchanger provides almost all of this outside air. The limited data available for sealed houses indicates that they typically experience winter infiltration rates of 0.1 to 0.2 air changes per hour. The ESSPS house has a still lower infiltration rate. The membrane that protects the house against infiltration is very tightly sealed, with fewer joints than the frame superinsulated house.

The walls are also protected by earth berms, which effectively block infiltration while creating a low profile for the building, so wind is deflected over the top of the house. Outside air infiltration will be in the range of 0.04 air changes per hour.

The rate at which ventilating air must be introduced into a sealed house is an important design issue. The first effectively sealed houses were constructed only a few years ago, and most of these buildings had serious habitability problems, because moisture and odors were trapped in the house during cold weather. Within a short time, a more esoteric enviromental hazard, Radon gas, was also found in sealed houses in many parts of the country. Radon gas, given off by uranium-bearing rock, has now been found to be present in significant concentrations in conventional homes in many areas of the United States and other countries. Radon's significance as a health hazard has not yet been fully evaluated, but it does seem clear that if a health hazard exists, sealing the house can make it more serious.

These problems have created debate over the minimum acceptable rate of air change in single-family dwellings. Rates as low as 0.2 changes per hour have been suggested, although a minimum standard of 0.5 air changes per hour has recently been established by Canadian authorities. The debate will only be resolved after prolonged experience with substantial numbers of sealed or partially sealed houses. Meanwhile, energy should not be conserved by lowering ventilation rates excessively, since both health and comfort are involved. The house analyzed in this chapter is quite large, and 0.4 air changes per hour are recommended; when the house is smaller than 1,600 square feet, 0.5 air changes per hour are necessary due to the greater concentration of moisture and odors.

Table 9–1. Winter Heat Loss for 1975 and 1985 Conventional Homes*

	1975 House		1985 House	
		Loss		Loss
Basement**	(Uninsulated)	246	(Uninsulated)	246
Walls	(R-11 insulation)	171	(R-16 insulation)	114
Windows & doors	(Single glass)	330	(Double glass)	147
Roof	(R-11 insulation)	83	(R-19 insulation)	50
Air infiltration	(1.0 air ch/hr)	320	(0.6 air ch/hr)	192
Total hourly loss/°F		1,150		749

* Hourly loss in Btu/°F for 2,000-sq-ft house.
** Basement loss includes above-grade basement wall area.

Table 9–2. Winter Heat Loss for the ESSPS House*

		Day	Night
Slab	(Insulated)	36	36
Walls	(R-25 insulation/earth shelter)	43	43
Roof	(R-57 insulation)	40	40
Windows & doors	(Double glass)	161	—
Skylights	(Double glazing)	17	—
Thermal shutters	(R-17 insulation)	—	20
Air infiltration	(0.04 air ch/hr)	16	16
Ventilating air**	(0.36 air ch/hr)	30	30
Total hourly loss/°F		343	185

* Hourly loss in Btu/°F for 2,000-sq-ft house.
** 80% recovery rate with heat exchanger.

Daily and Monthly Heat Loss

The performance of the ESSPS house with its shutters open is entirely different from its performance with nighttime shutters closed. Heat loss must therefore be evaluated on a daily, or twenty-four hour, basis.

The extent to which shutters are kept open or closed will vary among different households. If all the children are in school, and both parents are employed full or part time, the house will not be occupied during most of the weekday daylight hours. Some families will have the last person to leave the house close the shutters in bad weather; others will be less diligent. The energy-conscious family may also keep the house completely or partly shuttered on stormy days when the house is occupied, just as drapes and blinds are sometimes drawn on dark, gloomy winter days. These possibilities for energy savings are not taken into account here. It is assumed, instead, that all of the shutters will be kept open for ten hours during each winter day, regardless of weather conditions, or the extent to which the house is actually occupied.

The calculation of daily, as opposed to hourly, loss must also take into account the differential between day and night outside air temperature as well as the gain from internal heat sources.

Once transmission and infiltration losses are reduced through earth shelter and superinsulation, the magnitude of the internal gains generated within the building becomes very important. These gains are principally generated by appliances, lights, and occupants. Relatively little research has been done on internal gain in dwellings; estimates of total heat generated within a house occupied by a family of four vary from 50,000 to 125,000 Btu per day, with most authorities accepting estimates in the range of 50,000 to 75,000 Btu per day. The lowest accepted figure, 50,000 Btu per day, is used here. Ten thousand Btu per day are added for heat from an internally vented dryer, making the total estimated internal gain 60,000 Btu per day.

Energy consumption for winter heating also includes the electricity used to operate the heat exchanger. The exchanger will supply 0.36 air changes per hour to the living spaces. If the volume of the house is 16,000 cubic feet, 5,760 cubic feet per hour of fresh air are needed. This is equivalent to 96 cfm. A standard counterflow unit rated at 180 cfms for low-speed operation will run just over thirty minutes in each hour, or twelve hours per day. Typically, a current generation unit will use 150–200 watts of electricity to move 180 cfm of air, although units with comparable heat transfer efficiency show considerable variation in the amount of power needed to move the same volume of air. Two hundred watts of electricity is a low level of energy consumption, but residential electrical energy in the Philadelphia area costs approximately three times as much per Btu as energy derived from fossil fuels.

The exchanger energy consumption is:

$$(0.2) \times (12) = 2.4 \text{ kwh per day}$$

Since 1 kwh = 3,413 Btu, and each Btu of electrical energy costs three times as much as a fossil fuel Btu, the following equation can be written to determine the cost of operating the exchanger, expressed as fossil fuel energy usage:

$$2.4 \times 3{,}413 \times 3 = 24{,}574 \text{ Btu per day}$$

Table 9–4 summarizes monthly space heating energy consumption of conventional and ESSPS houses located in the Philadelphia area, taking into account exchanger energy use. The table shows ESSPS space-heating

Table 9–3. Daily Heat Loss—Average Outside Temperature 32° F

	1985 Conventional House	ESSPS House
Daytime loss (10 hours)	268,000	122,800
Nighttime loss (14 hours)	417,300	103,100
Internal gain	(60,000)	(60,000)
Total net daily loss (Btu)	625,300	165,900

Average daytime temperature—34.2° F.
Average nighttime temperature—30.2° F.
Indoor temperature—70° F.

Table 9–4. Monthly Heating Energy Use in Philadelphia During Heating Season*

		Conventional House	ESSPS House		
	Av. Temp. (F)	Net Heat Loss	Net Heat Loss	Exchanger Energy	Total ESSPS Energy
Oct.	55.7	6.2	0.7	—**	0.7
Nov.	44.3	12.1	2.7	0.8	3.5
Dec.	33.9	18.3	4.8	0.8	5.6
Jan.	32.3	19.2	5.1	0.8	5.9
Feb.	33.2	16.9	4.5	0.7	5.2
Mar.	41.0	14.9	3.6	0.8	4.4
Apr.	52.0	8.0	1.3	—**	1.3
Totals		95.6	22.7	3.9	26.6

* In MBtu.
** Exchanger does not operate.

demand reduced to less than 24 percent of the demand in the upgraded conventional house. With the energy consumed by the heat exchanger included, the ESSPS space-heating requirement increases to 28 percent of the energy required to heat the upgraded conventional house. Passive solar gain will make up nearly all of the heat loss from the ESSPS house in the winter months, although it cannot compensate for the energy consumed by the heat exchanger.

10

The Passive Solar Storage System

Passive Solar Gain

In order to simplify the discussion of passive gain, it will be assumed that all the glass faces south, with windows forming a horizontal band 4 feet 6 inches high. Analysis is also simplified by treating the skylights as if they are horizontal, rather than sloping approximately 9 degrees with 2/12 roof pitch. The 2,000-square-foot house has 375 square feet of glass area and 40 square feet of skylights. The simplified building will have virtually the same passive performance as the houses illustrated in figures 4–5 and 4–7.

Clear-day passive gain greatly exceeds the daily loss of heat under average temperature conditions; the comparison in Table 10–1 shows a very large difference between clear-day gain and daily loss.

Since there will be only a few clear days each month—usually only one or two entirely clear days—clear-day gain is much less significant than the average daily gain during the winter months. Table 10–2 shows that while the average daily passive gain is much lower than clear-day gain, it also exceeds average daily net heat loss.

Because average daily gain during the winter months exceeds average daily heat loss, the house will receive more passive energy than it requires to make up heat loss throughout the heating season. Radiation acquired on clear days or relatively clear days

Table 10–1. Clear-day Passive Gain and Average Daily Net Heat Loss—Philadelphia Area ESSPS House*

	Clear-day Gain**	Average Daily Net Heat Loss	Clear-day Gain as % of Average Daily Net Loss
Oct.	534,000	21,900	2,438
Nov.	541,400	91,100	594
Dec.	528,800	154,100	343
Jan.	549,100	163,800	335
Feb.	568,000	159,600	356
Mar.	522,800	117,100	446
Apr.	409,300	44,400	922

* In Btu.

** Transmission for glass and skylights—0.81.

cannot, however, compensate for lack of gain during prolonged periods of bad weather unless a very large amount of storage capacity is installed. Daily variation in solar input must, therefore, be studied in order to determine the number of days during which daily gain will not be able to offset average heat loss. The *Solar Radiation Energy Resource Atlas of the United States* (Solar Energy Research Institute, Golden,

Table 10–2. Average Daily Passive Gain—Philadelphia Area ESSPS House

	Average Daily Passive Gain*	Average Daily Passive Gain as % of Average Daily Net Heat Loss
Oct.	293,900	1,340
Nov.	261,900	287
Dec.	230,500	150
Jan.	251,800	154
Feb.	293,800	184
Mar.	305,300	261
Apr.	316,100	712

* In Btu.

SOURCE: Solar Radiation Energy Resource Atlas of the United States.

Table 10–3. Monthly Passive Solar Deficits*

	No. Deficit Days	Average Size of Daily Deficit**	Monthly Deficit
Nov.	6	37,000	222,000
Dec.	11	83,900	923,000
Jan.	11	87,800	966,000
Feb.	9	83,700	753,000
Mar.	6	59,100	355,000

* In Btu.
** Average daily heating requirement minus average daily gain during deficit days.

Colorado, 1981) gives the daily distribution of solar energy levels on south-facing surfaces by months of the year for twenty-six locations in the United States. This data can be used to obtain the number of deficit days—the days that passive gain does not equal average daily loss—in each month. The data can also be used to obtain the size of the average energy shortfall on deficit days, so the size of the monthly passive energy deficit can be calculated. This deficit must be made up during the month either by carrying

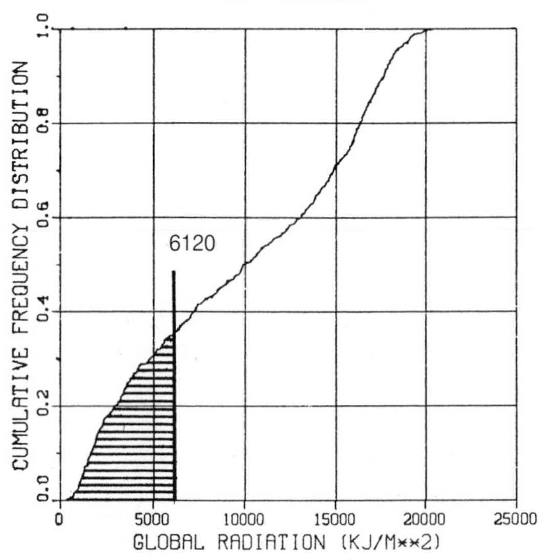

10–1. Daily variation in solar radiation (January: Sterling, VA). The chart shows daily variation in solar radiation in January (global radiation on a south-facing surface). When radiation exceeds 6,120 KJ/M² (539 Btu/ft²), enough energy is received to heat the Philadelphia area ESSPS house at average January temperatures. Days on which less than 539 Btu/ft² of radiation are received on a south-facing surface are deficit days. The shaded area on the chart gives the size of the monthly passive deficit.

over stored passive energy to deficit days, or by using the backup heating system.

The distribution chart for January (using the reporting station closest to Philadelphia with comparable weather patterns) shows eleven deficit days (figure 10–1). The average size of the deficit is 87,000 Btu, compared with average January daily heat loss of 163,000 Btu. The chart shows only a few days during the month when passive gain is negligible and indicates that significant gain is usually obtained through cloud cover. The passive gain received by the heavily protected ESSPS house during deficit days—days with the poorest weather—offsets nearly half the average daily heat loss. This greatly reduces the need for storage to compensate for bad weather.

Storage Design

Storage capacity in the ESSPS house is provided by the floor slab and drywall or plaster walls and ceilings. Studs and ceiling rafters, as well as furniture and other interior fittings, also retain passive gain, but their contribution to building storage capacity is relatively small.

The choice of floor materials has an important effect on slab storage capacity, and slab performance is analyzed here with vinyl flooring in the kitchen, family room, dining room, and atrium-gallery adjacent to these spaces. The rest of the living area is carpeted; the utility room slab is left exposed. Slab thickness is increased to 6 inches where vinyl is laid, or where the concrete is not covered. With these floor coverings, the slab, walls, and ceilings provide half the total storage capacity.

All of the slab can be used for storage, because the fan and ducts of the backup heating system circulate air throughout the house when passive energy is being acquired. A maximum air circulation rate of 1,000 cfm is required to distribute heat acquired during clear-day peak periods to the entire slab. When passive gain is discharged mechanical air circulation is also desirable in order to maintain even temperatures in the building.

The required rate of air circulation is much lower during discharge periods than during the peak acquisition periods. The amount of electricity used by the fan motor of the heater for circulation of passive energy is small and will be the same or less than the amount used by the furnace fan of a conventional house equipped with a warm-air heating system.

Additional storage capacity is provided by 66 lineal feet of 8-inch block walls that are built as partitions between the bedrooms. Tubes of Glauber's salt (phase change materials) are placed in a plenum located in the mechanical room and attached to the ducts of the backup heating system. The air-handling unit of the backup system provides a rapid flow of air around the tubes; this air flow is needed to use the full capacity of the phase change materials.

A simple method for calculating the capacity of the mass storage components of a passive system is given in the *Passive Solar Design Handbook* (New York: Van Nostrand Reinhold Company, 1984). This method assumes a twenty-four hour or diurnal storage cycle. During this cycle, temperature in the house is allowed to fluctuate freely under passive load, and all of the heat stored by the system is also assumed to be given up at the end of the diurnal cycle.

The free sinusoidal temperature swing does not correspond to actual conditions in the ESSPS house, where a maximum temperature swing of 8° F is permitted (67° F to 75° F). Maximum temperature of the storage elements will be achieved partway through the storage phase of the cycle and maintained until the end of the input period. Because the rate of heat transfer from the room to the cooler storage mass depends on the room air temperature, application of the diurnal method, with its freely fluctuating temperatures, will underestimate the storage capacity of the ESSPS house (figure 10–2).

Heat storage capacity for the components of the passive system is calculated in Appendix B. The results are summarized in Table 10–4.

Table 10–4. Daily Storage Capacity of the ESSPS Passive System*

Storage Element	Storage Capacity
Gypsum board	22,700
Slab (vinyl-covered/exposed —1,150 sq. ft.)	39,700
Slab (carpet—1,330 sq. ft.)	19,600
Block storage walls (66 lineal feet)	31,000
Phase change material	52,000
Total capacity	165,000

* In Btu.

Passive Solar Storage and Winter Heating Demand

At a minimum, a storage system must be able to keep the house warm overnight in very cold weather. The outside winter temperature for the Philadelphia area—the coldest temperature that will normally be encountered—is 0° F. The ESSPS house is heated to 75° F as its storage system is

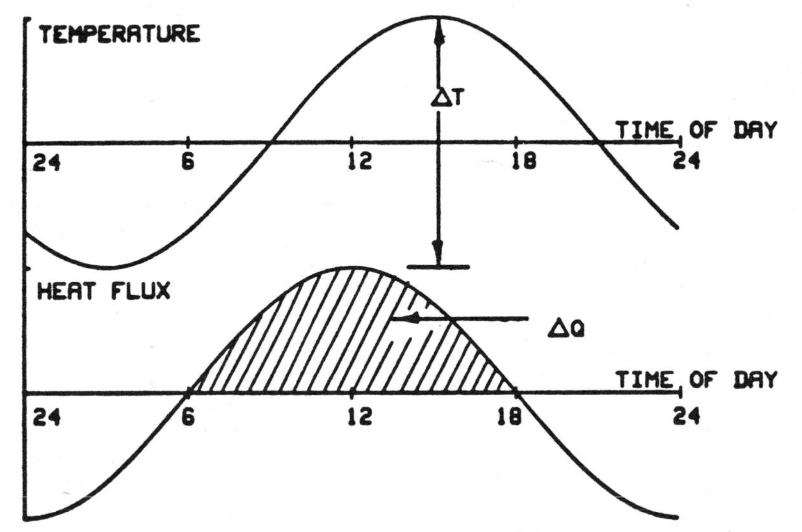

10–2. Daily temperature fluctuation under passive solar load. The graph shows a sinusoidal temperature-change cycle assumed by the diurnal method.

filled, and if temperature inside the house is not to drop below 67° F, average temperature during the night will be approximately 71° F. Heat loss for the night will be

$$185 \times 71 \times 14 - 30{,}000 = 154{,}000 \text{ Btu}$$

where:		
	heat loss/°F with shutters closed	= 185 Btu/hr/°F (Table 9–2)
	duration of nighttime period	= 14 hours
	nighttime internal gain	= 30,000 Btu

The storage system, with a capacity of 165,000 Btu, will be more than adequate to keep the house warm through the coldest nights of the winter.

If average nighttime temperatures during the winter months are considered, rather than the coldest nights, the passive system will take the building through until morning and retain over 100,000 Btu of stored heat. This heat can be applied, when necessary, to the deficit days.

The length of time that the house can be heated by the storage system during deficit periods will depend on the amount of energy received on deficit days to supplement stored heat obtained during the preceeding surplus days. Average net twenty-four-hour loss during the coldest months is approximately 160,000 Btu. On the rare days when passive input is close to zero, the storage system will only be able to supply about sixteen hours heat, in addition to supplying average nighttime heat loss. But with average energy deficits, the storage system is able to supply heat longer. Average daily deficits during the winter months in the Philadelphia area are only 85,000 Btu. Stored heat in the system will meet demands over one and a quarter average deficit days in December, January, and February. In March the system will be able to meet requirements for two full successive average deficit days; in November it will heat the house for three and one-half successive deficit days.

The pattern of alternating surplus and deficit passive input days thus becomes very significant. If a week of surplus days is regularly followed by three or four deficit days, carryover in the storage system will be applied only three times each month, and the storage capability will be periodically exhausted. If each deficit day is preceded by a surplus day, all daily deficits can be made up from storage, provided there are no large deviations from the monthly average temperature. For a full picture of the performance of the passive system it is necessary to have accurate data for the pattern of insolation (solar gain) during the month, as well as data that correlates daily temperatures with solar gain.

An adequate data base for a valid model of this type has not yet been developed. Some indications of the general characteristics of the weather patterns that the complete model would show are, however, available. Cloudy days in the Philadelphia area typically are slightly warmer than clear days, indicating that average temperature levels for the deficit days will be somewhat above the monthly average. Limited data for insolation fluctuations on the east coast, published in the *Solar Radiation Atlas* (Solar Energy Research Institute, Golden, Colorado, 1981), also indicate that the general pattern of radiation variation will be an alternation of one, two, and three successive surplus and deficit days (Appendix C).

Under these conditions, energy stored in the passive system will be used at least five times each month, providing 40,000 to 125,000 Btu on each occasion. In November and March, the storage system will be able to meet deficits during repeated periods of bad weather, lasting from two to four days. Passive capacity will only be exhausted during very rare periods of extended bad weather, or when an unusual combination of

Table 10–5. Nighttime Heat Loss and Passive Solar Storage Capacity*

	Average Nighttime Outdoor Temp.**	Average Nighttime Heat Loss	Average Nighttime Net Heat Loss***	Average Stored Heat at 7 A.M.****
Nov.	42.5	64,000	34,000	131,200
Dec.	32.1	86,500	56,500	108,700
Jan.	30.5	90,000	60,000	105,200
Feb.	31.4	88,000	58,000	107,200
Mar.	40.4	68,600	38,600	126,600

* In Btu.
** Philadelphia area.
*** 30,000 Btu nighttime internal gain.
**** Total storage capacity is 165,000 Btu.

exceptionally heavy cloud cover and below-average temperatures occurs. With normal weather patterns, the storage system will be able to supply all of the November and March passive deficits. During the three coldest months of the season, storage capacity is exhausted after two average deficit days, and the backup system is used. Total energy supplied by the backup system for the Philadelphia area house will, however, be very small.

Table 10–6. Monthly Passive Deficit and Estimated Offset From Storage*

	Monthly Passive Deficit	Energy Supplied by Storage System During Deficit Days	Remaining Energy Deficit
Oct.	—	—	—
Nov.	222,000	222,000	—
Dec.	923,000	540,000	383,000
Jan.	966,000	525,000	441,000
Feb.	753,000	450,000	303,000
Mar.	355,000	300,000	55,000
Apr.	—	—	—
Totals	3,219,000	2,037,000	1,182,000

* In Btu.

Table 10–7. Winter Space-heating Energy Use for Philadelphia Area ESSPS House*

Conventional House	ESSPS House			
Net Heat Loss	Net Heat Loss	Heat Exchanger Energy Use	Passive Input	Total Heating Energy
95.6	22.7	3.9	21.5	5.1

* In MBtu.

11

Designing for Summer Cooling

Minimizing Summer Cooling Loads

The heavy thermal envelope of the ESSPS house reduces transmission gain during hot weather, just as it reduces transmission gain in winter. The earth sheltered walls of the house are thermally neutral in summer—they do not contribute to heat gain; gain through the roof is also reduced. But the extra protection provided by superinsulation and earth shelter does not greatly reduce the total cooling load. This is simply because in a moderate climate such as the Philadelphia area, a relatively small proportion of the cooling load is generated by transmission gain through walls and roof. In Table 11–1, where gain is calculated for an outside temperature of 85° F, roof and wall transmission account for only 16 percent of the total clear-day load. Windows are much more important as is heat gain from internal sources and air infiltration.

Shielding the windows of the ESSPS house from direct radiation has a very significant effect on cooling load. In Table 11–1 the conventional house is assumed to have windows of equal size facing north, south, east, and west, with half of each window protected by light-colored blinds or drapes. Those of the ESSPS house face south and are protected from direct radiation by sunscreens, which, however, do little to reduce the amount of indirect radiation that reaches the windows. Indirect radiation—solar radiation that is diffused by cloud cover or dust particles in the atmosphere—reaches the building at all angles from horizontal to vertical (figure 11–1) and is commonly estimated to make up one-third of the total clear-day solar radiation that falls on a south-facing surface. Protection against indirect radiation must be provided by drapes or blinds; in Table 11–1 the windows of the ESSPS house are assumed to be half covered during the day.

Much less can be done to limit internal heat gain and gain from air infiltration and ventilating air. Internal heat generation can be reduced by venting the refrigerator, oven, and dishwasher, as well as by tightening bath and shower enclosures. This is estimated to lower internal gain by 15 percent (7,500 Btu per day). The table shows no decrease in heat gain from outside air. Ventilating air requirements in summer are the same as in winter—0.4 air changes per hour. Summer infiltration rates for a conventional house are customarily estimated at about 0.6 air changes per hour (ASHVE Guide, 1981), but with a partial infiltration barrier such as

Table 11–1. Air-conditioning Loads at 85° F*

Conventional House	Load (Btu/hr)
Walls	1,800
Roof	1,700
Glass	11,300
Infiltration**	1,100
Latent Heat	3,200
Internal Sources	2,500
Total Load	21,600
ESSPS House	
Walls	100
Roof	1,600
Glass and Skylights	6,200
Infiltration*	1,100
Latent Heat	3,200
Internal Sources	2,100
Total Load	14,300

* Indoor Temperature—77° F.
** Sensible Heat. Adjusted for 0.4 air ch/hr.

SOURCE: ASHRAE Guide, 1981: Simplified Method for Residential Air Conditioning Loads.

Table 11–2. Energy Costs for the Philadelphia Area Conventional and ESSPS House

	Conventional House	ESSPS House
Space Heating*	$ 912	$ 41
Air Conditioning**	154	87
Hot Water Heating***	338	202
Energy Cost****	$1,404	$330

* Fuel Oil: $1.05/gal.
** Electricity: $0.0939/kwh (Source: Philadelphia Electric Co.)
*** Electricity: $0.0583/kwh (off-peak rate) (Source: Philadelphia Electric Co.)
**** Excluding other electrical usage.

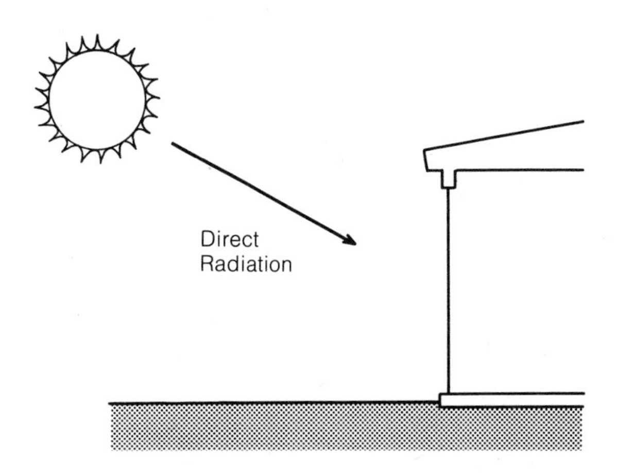

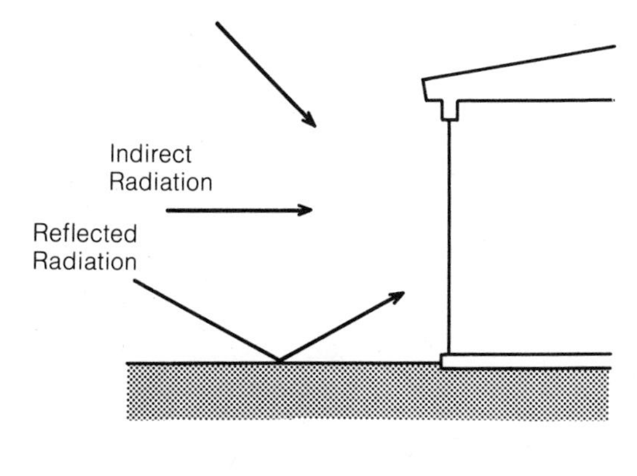

11–1. Solar radiation.

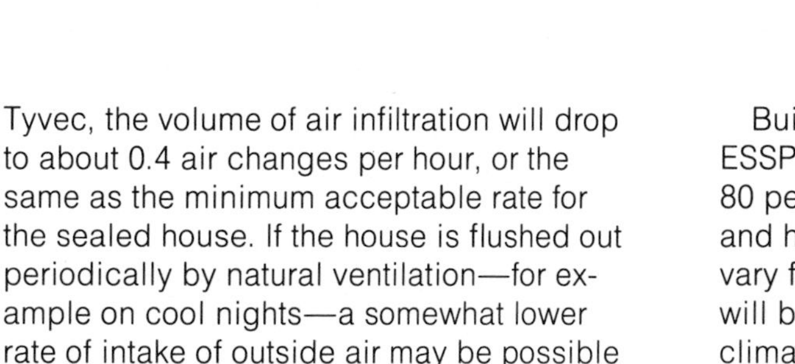

Tyvec, the volume of air infiltration will drop to about 0.4 air changes per hour, or the same as the minimum acceptable rate for the sealed house. If the house is flushed out periodically by natural ventilation—for example on cool nights—a somewhat lower rate of intake of outside air may be possible when the house is sealed during the warm hours of the day.

The large savings in transmission gain through the shielded windows, combined with smaller savings in heat gain through the walls and roof and from internal sources, reduce cooling loads on clear days by one-third. A further improvement in the hot-weather performance of the ESSPS house is obtained from the hot-water preheat system. The preheat system absorbs heat from the room air that circulates through its heat exchanger. If an average of 75 gallons of water per day are preheated from 50° F to 78° F, the preheat system will absorb 17,550 Btu per day, and cooling loads are reduced by a total of almost 45 percent.

Built in southeastern Pennsylvania, the ESSPS house produces a savings of almost 80 percent on the cost of heating, cooling, and hot-water supply. Energy savings will vary for different regions of the country and will be substantially larger in more severe climates.

Modifications for Milder Climates

In regions with heating seasons in the 3,000 to 5,000 degree-day range, the building design should be modified to achieve a still greater reduction in summer cooling loads. The inner portion of the slab can be left uninsulated, allowing it to become a heat sink in summer. Winter passive gain can also be sacrificed in a milder climate; smaller windows with triple glass are appropriate. If window area is reduced from 375 to 300 square feet and skylight area from 40 to 25 square feet, using a third layer of glass, the cooling load will be reduced by over 15 percent; the cooling effect of the uninsulated inner slab must be added to this substantial savings.

Landscaping can also be used in any climate to reduce cooling loads. Ground reflectance provides a large share of the direct radiation reaching the south-facing glass, and reflected radiation cannot be blocked by sunscreens. This radiation represents at least 20 percent of the total radiation reaching a south-facing vertical surface near ground level. If the berms under the windows of the ESSPS house are covered with grass, the house will receive a large amount of reflected radiation—reflection from grass is very high, approximately equal to that of asphalt paving. Planting berms with darker color ground cover will decrease the amount of reflected radiation. It is also important to keep the deck or patio at least 10 feet away from the house, so its reflective surface does not direct sunlight against the windows.

Deciduous trees, as well as plantings, can be used to control indirect radiation. If the trees are planted as close as possible to the house, their branches will block a minimum of winter sunlight, with branches of mature trees interrupting the sunlight only in late winter and early spring (February 15 to April 1). Once in leaf, the trees help limit the amount of indirect radiation reaching the windows and will shade part of the house and garden.

PART IV

CONSTRUCTION COST AND COST-EFFECTIVENESS

12

Analyzing Cost-effectiveness

Variables in the Analysis

The ESSPS house reduces the winter heating bill by over 90 percent and saves nearly half the cost of hot-water heating and summer cooling. These savings in energy consumption are extremely impressive, but the crucial question is whether the energy-efficient design is cost-effective. A decrease in energy consumption and energy costs must also reduce the overall cost of owning a home.

The equation that determines the cost-effectiveness of an energy-efficient house includes both energy consumption levels and energy prices. Other factors include construction costs, mortgage rates, tax deductions, and maintenance costs for the building and its equipment. These factors are variables: some change over time, others according to location.

Energy prices and construction costs have risen rapidly in recent years. Mortgage rates, which transfer construction costs into the cost of owning a home, have been just as volatile, and the 5 percent and 6 percent rates of the late 1960s have given way in recent years to rates of 12 percent and higher. Tax rates and tax credits, although changing less frequently than the other variables, are also subject to change as tax rates are altered and special tax provisions for the energy-efficient house are revised.

Building costs and energy prices, while subject to broad changes over time, are also subject to significant local and regional variation. Mortgage rates show much lower regional variation, and tax rates are uniform throughout the United States; cost-effectiveness must, however, be evaluated on the basis of an entirely different tax structure for homes built in other countries. Energy consumption does not vary over time, but more than any other single factor, it is subject to local and regional variation.

The balance sheet for the cost-effectiveness of an ESSPS house must be drawn for a specific building in a particular location and, unlike the energy budget, it will be valid only at the time it is drawn.

Concepts of Cost-effectiveness

Cost-effectiveness can be defined in many ways, and conflicting standards of cost-effectiveness have been used by advocates of the various types of energy-efficient homes. The familiar payback concept was originated by advocates of active solar, who recognized that the active system represented a large investment that would have to be paid back over an extended period of time. Implicit in most payback analyses, particularly those developed for active solar, is the assumption that energy prices will continue to rise rapidly, so the payback analysis is usually oriented toward a future with much higher energy prices than those that prevail when the house is built.

The development of superinsulation has led to the formulation of a very different concept of cost-effectiveness, one that focuses on the present, ignoring possible escalation of energy costs in the future. Superinsulation is regarded as cost-effective if the additional cost of the house, as represented by increased annual mortgage payments, with tax credits and other offsetting factors taken into account, is smaller than immediate savings in fuel costs. When this method is used, each part of the design can be evaluated separately to determine whether it provides immediate benefits. The procedure allows an "optimum" amount of insulation to be calculated, and the economic feasibility of other features, such as the heat exchanger, is also determined on the basis of their cost and the current fuel savings that they provide. This method maximizes current benefits from the energy-efficient building, since it can be used to exclude any component that will not generate immediate savings.

The concept of cost-effectiveness that has been used in designing the ESSPS house seeks a middle ground. It places heavy emphasis on present conditions while recognizing the strong possibility of escalating energy prices. The cost-effectiveness of the energy-saving features of the ESSPS house are considered together, rather than separately; if a balance is created between increased mortgage payments (after tax credits and other offsetting factors have been taken into account) and current savings in energy costs, the house is considered cost-effective.

This method allows features that do not pass the test of immediate cost-effectiveness to be included—the buildings will have more insulation and include energy-saving features that are not incorporated in a house designed to maximize current benefits. The result is that at first the ESSPS house does not yield significant net savings, but in the long run, if energy prices increase, whether slowly, or in time of crisis, very rapidly, it will save a great deal more money than a house designed to create the largest possible immediate cost benefits.

The ESSPS house can also be designed to fit the requirements of both of the other concepts of cost-effectiveness. If the standard of maximizing immediate savings is adopted, less insulation will be used, the heat exchanger will probably be replaced with simple mechanical ventilation, and the hot-water preheat system will be less elaborate. If continued rapid long-term escalation of energy costs is considered certain, more insulation will be added, passive storage capacity will be increased, and the hot-water heating system will be further refined. Since events in the future determine the validity of these choices, the extent to which energy-saving features are incorporated in the ESSPS house must be left to the individual designer and homeowner.

13

Energy-efficient Design and Construction Costs

Comparative Construction Costs

Construction costs of the energy-efficient house are central to the cost-effectiveness equation regardless of how cost-effectiveness is measured. The ESSPS house will cost more to build than the conventional house, because it includes more insulation as well as special energy-saving features such as its passive solar storage system, thermal shutters, and heat exchanger. Comparative construction costs for the ESSPS house and a conventional tract house are shown in Table 13–1. This table includes the builder or general contractor's raw costs—payments to subcontractors and the cost of materials and labor directly purchased by the builder or contractor. Overhead, profit, and selling expense are taken into account in Table 14–1.

Some components of the building—electrical work and plumbing, kitchen and bathroom equipment, interior finish (except window trim), floor coverings, and painting—are little changed and need not be part of the cost comparison. Neither is the cost of the lot usually affected by energy-efficient construction.

In order to make the cost comparison realistic, the basic ESSPS house (see figure 4–6) is measured in Table 13–1 against a modest tract house with the same amount of living space. This house is a plain, two-story box with a full basement. This type of house sold in the Philadelphia metropolitan area in 1985 for $110,000. The building has aluminum siding on its exterior and its outside is embellished only by a brick veneer front. It is, however, equipped with the full range of current energy-saving improvements, including heavier insulation, good quality wood windows with insulating glass, and a partial infiltration barrier of Tyvec.

The basic ESSPS house is also simply designed. Interior walls are of gypsum board, with painted block storage walls. Features that add extra interest to this simple interior include a 6-foot-wide atrium with planters and a continuous trellis under the skylights, stained mahogany trim, and a change in level at the living room end of the house. Exterior materials are brick for exposed house walls and retaining walls, and redwood or cedar for trim. The cost of both houses could be reduced by approximately the same amount if the brick veneer were eliminated on the front of the conventional house, so that it has an all aluminum exterior, while the brick used on the ESSPS house for above-grade walls is replaced with stucco .

Shell Cost

The cost of the shell of the ESSPS house is 10 percent higher than that of the conventional house. This differential is almost exactly the same as the difference in insulation cost. Most of the other major parts of the shell also show significant cost differences, but these differences largely offset one another.

Masonry costs shown in Table 13–1 include horizontal reinforcing of earth sheltered walls and piers at 16-foot intervals. The retaining walls represent a substantial addition to masonry costs, but this is offset by the lower cost of chimney construction. The chimney of the earth sheltered house is only 1½ stories high, compared to 3½ stories for a two-story house with basement, and most of its surface is below grade, so much less brickwork is required. Construction is also much easier, because masons can stand on the earth berms and the roof of the house to complete the chimney; they will not need the elaborate scaffold required for constructing a taller chimney. Total masonry costs for the ESSPS house are higher even with savings on chimney construction, but this is offset by lower framing costs. Thirty percent less lumber is needed for the ESSPS house and the framing operation, carried out on the slab, is very efficient.

Insulation used in the ESSPS house to form its heavy thermal envelope costs three times as much as insulating the conventional house. Insulation is inexpensive, however, and the batt insulation and polystyrene sheathing of the conventional house account for slightly over 2 percent of its construction cost. In the ESSPS house insulation represents just over 7 percent of construction cost, and as the single most important en-

Table 13–1. Construction Cost Comparison*

Shell	ESSPS House		Conventional House	
Masonry	Footings	$ 1,250	Footings	$ 650
	Slab (4″ stone)	3,800	Slab	1,400
	Block walls	4,000	Block walls	2,800
	Footing drain	450	Waterproofing	50
	Waterproofing	350	Fireplace & chimney	2,750
	Fireplace & chimney	1,500	Brick veneer	4,700
	Brick veneer	3,800		
	Retaining walls	1,500		
		$16,650		**$12,350**
Framing lumber	Walls	$ 1,400	Steel	$ 400
	Roof	3,450	Walls	2,450
	Membrane	100	Floors	2,750
			Roof	1,700
		$ 4,950		**$ 7,300**
Carpentry labor	Framing	**$ 3,250**	Framing	**$ 4,700**
Insulation	Slab	$ 850	Walls**	$ 1,400
	Walls	1,600	Roof	450
	Roof	3,350		
		$ 5,800		**$ 1,850**
Windows/doors/ skylights	Casement sash	$ 1,600	Windows	$ 5,250
	Fixed glass	2,200	Window trim	600
	Window trim	1,950	Patio door	450
	Entrance doors	600	Entrance doors	450
	Garden door	200		
	Skylights	600		
		$ 7,150		**$ 6,750**
Trellis (Gallery)		**$ 500**	**Stairs**	**$ 900**
Exterior	Roofing	$ 2,250	Roofing	$ 1,200
	Barge & soffits	350	Tyvec membrane	250
			Aluminum siding	2,300
			Barge boards	150
		$ 2,600		**$ 3,900**
Total Cost of Shell		**$40,900**		**$37,750**

Table 13–1. Construction Cost Comparison continued.

Equipment	ESSPS House		Conventional House	
Mechanical	Backup heating & air-conditioning	$ 3,300	Heating & air-conditioning	$ 4,800
	Hot water	2,200	Hot water	200
	Heat exchanger	1,500		
Thermal shutters		**1,800**		**$ 5,000**
Sun shades		**1,000**		
Total Equipment Cost		**$ 9,800**		
Passive Storage				
	Block walls	$ 1,100		
	Thickened slab	300		
	Phase change materials	1,100		
Total Storage System Cost		**$ 2,500**		
Sitework & landscaping				
	Excavating & grading	$ 1,900	Excavating & grading	$ 1,500
	Walks & deck	900	Walls & deck	700
	Landscaping	2,100	Landscaping	700
Total Sitework & Landscaping		**$ 4,900**		**$ 2,900**
Total Construction Costs		**$58,100**		**$45,650**

* Prices for 2,000-sq.-ft. house.
** Includes 1" polystyrene sheathing.

ergy-saving feature of the house, its cost is very low.

Equipment and Passive Storage

Equipment costs show a wider differential than shell costs, because these include, for the ESSPS house, thermal shutters and summer sunscreens, as well as the heat exchanger and hot-water preheat system. Both houses are equipped with central-heating and air-conditioning systems designed to the same performance standards, but equipment capacity and duct sizes are smaller for the ESSPS house. Additional savings are obtained by installing a heat pump in the ESSPS house. The heat pump costs much less to install than a combination of an oil or gas furnace and separate air-conditioning equipment, but it is more expensive to operate in cold climates, since its efficiency falls sharply below 40° F. Using the heat pump is practical in the ESSPS house, because the backup heating system only operates for short periods during cold weather.

The cost of the passive solar storage system, the third category in Table 13–1, is relatively small, because the requirement for storage capacity is low, and building mass fulfills a large share of this requirement. The reduced cost of passive storage is a very important element in the overall cost-effectiveness of the ESSPS house. Extra construction for passive storage includes a 6-inch slab wherever the floor is not carpeted. This addition to the building is comparatively inexpensive since no extra labor is required; the additional concrete is the only added expense. Block storage walls and phase change materials provide somewhat more expensive capacity. Table 13–1 does not include plaster or other decorative finishes for the storage walls, but it allows for stack-bonded construction and carefully finished joints.

Excavating, Grading, and Landscaping

Excavating and grading expense for the ESSPS house will be only a few hundred dollars higher than for a conventional house. The cost increase is held down because the number of operations requiring heavy equipment does not change. Earth sheltered design will usually require moving a great deal more earth to form berms, or to cut down banks, but this will not involve a large amount of machine time. A 2-yard, front end loader will shape the berms around the house and those planned elsewhere on the site in one day, whereas a medium-size, 1⅓ yard, machine will typically require a day and a half.

Although excavating and grading work will not be expensive, the landscaping budget for the ESSPS house is very substantial, reflecting the importance of landscaping in the design of the building.

The basic landscaping work should include ground cover on banks surrounding the house, a number of large bushes or small ornamental trees, several large shade tree saplings, and establishment of the lawn in areas where ground cover is not planted. The walks are of particular importance, and money is budgeted for ornamental walks of brick or flagstone, although the deck is built of treated framing lumber. A redwood deck or a patio of brick, stone, or decorative concrete is preferable, but this will stretch the builder's budget unless offered as an extra. All of the basic landscaping work must be regarded as part of the construction cost of the ESSPS house, since without it the building will seem bare and incomplete until the homeowner has made a substantial additional investment in the property.

14

The Balance Sheet

Sale Price and Mortgage Payments

Table 13–1 shows the builder's raw costs. Overhead and profit must be added to raw costs, and construction financing must also be included. If the house is built for sale, sales commission must be added as well as, in some states, real estate transfer tax.

Table 14–1 shows a raw construction cost differential of $12,450, producing a difference of $15,180 in the sale prices. If the conventional tract house sells for $110,000, the comparable ESSPS house will have a price of just over $125,000. The price differential will be approximately the same for a house built by an individual. Here the differential will include increased allowance for the general contractor's overhead and profit, increased construction financing charges, and larger architectural and engineering fees.

With a higher sale price, or higher total cost for the home built by an individual, the energy-efficient house will require a larger mortgage. The bank will be willing to provide the bigger loan, since the loan officer's analysis of total carrying costs will show lower energy bills to offset the increased interest and amortization payments for the loan. The level of mortgage interest rates will determine the size of the addition to annual carrying costs. If the lending rate is 12 percent, and the mortgage period twenty-five years, the increase in yearly payments for the basic ESSPS house will be $1,920. With 10 percent interest, the increase falls to $1,656, and at 8 percent to $1,406. These amounts are considerably larger than the en-

Table 14–1. Sale Price Differential—ESSPS and Conventional House

	ESSPS House	Conventional House
Construction costs, materials & labor*	$58,100	$45,650
Builder's overhead & profit (12%) and Financing costs (4%)	9,300	7,300
Total construction cost	67,400	52,950
Real estate commission (4%) and tax** (1%)	3,400	2,670
Partial Sale Price	$70,800	$55,620

* Items scheduled in Table 14–1.
** Pa. real estate transfer tax.

ergy savings of $1,100 projected for the Philadelphia area house, but all or most of the differential between energy savings and extra interest and amortization charges is offset by federal tax credits.

Federal Tax Credits

Most of the mortgage payment for a newly built house goes for interest; very little is applied to amortization. Interest paid on a home mortgage is deductible from taxable income. If the family is in the 25 percent bracket, the first-year credit, with 12 percent interest, is $454 and the after-tax addition to carrying costs is only $1,466 compared with energy cost savings of $1,100.

Application of the renewable energy tax credit also has a significant effect on the balance sheet of the ESSPS house. Enacted in 1978, initially to cover the tax years 1977–85, this credit allows a tax deduction of 40 percent on a maximum of $10,000 of eligible items included in the house, so that there is a maximum tax credit of $4,000. The renewable energy credit was originally enacted to promote use of active solar, but it was amended to cover passive solar. Items that are part of the structure of the house were, however, specifically excluded from eligibility for credits, and the energy deduction has not been revised to extend full benefits to the newer superinsulation concept.

Under present (1985) law, credits can be taken for the passive storage system of the ESSPS house, for the heat exchanger, and for the hot-water preheat system. The builder's raw cost for the eligible items is $6,000, and the prorated portion of the sale price or total cost of the house is $7,300. This creates a tax credit of slightly over $2,900. The credit can be taken during the first year of ownership to reduce the mortgage. After the tax credit is applied, the extra mortgage balance is reduced to just over $12,000, and added annual payments drop to $1,550. The credit for the interest portion of this payment is $367, and the after-tax addition to carrying costs is $1,183, or very nearly the same as the amount saved on energy costs.

The "bottom line" of the balance sheet will be altered by changes in the relationship between construction costs, interest rates, and energy prices, as well as by future tax legislation. Lower interest rates will have a strongly favorable effect. Higher construction costs, typical of most major metropolitan areas in the United States, will unfavorably affect the ESSPS balance sheet. On the other hand, energy savings will also be higher in many areas. Philadelphia's heating season of 5,100 degree days is not severe for the northern part of the United States, where most metropolitan areas experience 5,500 to 7,000 degree days, while a few have as many as 8,000.

Table 14–2. The Effect of Federal Tax Credits

Differential sale price	$15,180
Estimated renewable energy credit (1985 law)	2,920
Increase in mortgage after tax credit	12,260
Increase in annual mortgage payments (12% 25-year loan)	1,550
Federal income tax credit for interest payments* (first year of loan)	367
Net extra payment (first year)	1,183
Energy savings	1,074
Extra cost of owning ESSPS house (first year)	$ 109

* 25% federal income tax bracket.

Long-term Costs and the Effects of Inflation

Built in most parts of the central and northern United States, the ESSPS house will show either nominal extra expense or produce relatively small savings during the first year following construction. Cost-effectiveness in later years will depend primarily on the effects of inflation.

In a fully noninflationary economy, the ESSPS house represents an investment for the future. It offers security against sudden increases in energy costs, but should these not occur, large cost benefits will be realized once the mortgage has been paid off. If the house is financed in a noninflationary economy at a rate of interest comparable to present levels, the energy-saving features will not quite pay for themselves while the mortgage is amortized (Appendix D). But in a stable, noninflationary economy, interest rates would be in the range of 4 to 6 percent, or even lower. The lower rates would produce a favorable cash balance while the house is being amortized; after the mortgage has been paid off the full effects of the energy savings are realized (Appendix D).

Obtaining real cost benefits from a fully paid mortgage (or its equivalent, if the house passes through the hands of a number of homeowners by resale) requires low long-term maintenance and equipment replacement costs. The ESSPS house is designed for minimum exterior maintenance and has a much sturdier structure than the conventional frame dwelling. Most of its energy-saving features do not require replacement, and long-term maintenance costs associated with the energy-saving design include only repair and replacement of the fan motors of the heat exchanger, replacement of the hot-water heating system, and renewal of the phase change storage tubes located in the main duct plenum.

Replacement of the fan motors of the heat exchanger will be necessary at ten- to twenty-year intervals. The expense for this is not large; the motors cost only a few hundred dollars. Replacement of hot-water heating equipment is also necessary after ten to twenty years. For the ESSPS house, this will mean replacement of both its conventional heater and the preheat equipment, including the storage tank, air-to-water heat exchanger, and the motor of the air-handling unit. Tubes holding the phase change storage material will also require replacement at intervals that its manufacturers estimate at twenty years, but which may prove to be shorter.

These modest replacement costs are largely offset by the much longer service life of the backup heating equipment, which will operate for only a few days each year. The very small extra expense for replacement of mechanical equipment, combined with low maintenance cost for the house itself make realization of long-term cost benefits realistic in the noninflationary economy.

When the effects of inflation are considered, the ESSPS house provides a much higher degree of cost-effectiveness. The United States has experienced steady inflation for over forty years, and prices during this period have risen at annual rates of 3 to over 10 percent. This has made its home the American family's best investment, worth, within a short time after it is built, a great deal more than the purchase price. The energy-efficient house extends the beneficial effects of inflation to the homeowner. More money is invested in the house, to create a saving in energy costs. Once this investment is made, the burden of the extra mortgage payments diminishes rapidly in an inflationary economy, while the savings in energy costs increase from year to year.

The future course of energy prices cannot be charted with any degree of accuracy. In the long run, the cost of energy will rise at a rate at least equal to the rate of worldwide inflation and will probably exceed this rate. Recent changes in the prices of individual fuels have been extremely dramatic. During the past few years, prices of electricity and natural gas have risen very rapidly in the United States, with annual increases, in many areas, of 15 to 20 percent, while the cost of fuel oil has dropped from the peak that it reached after doubling in price between 1979 and 1981. Stable costs for fuels from domestic sources appear, in contrast, to be unlikely in the near future. Natural gas supplies are increasingly dependent on very deep wells, a more costly source of supply than shallow wells, and long-term gas price levels will also be affected by rising drilling and exploration costs.

The outlook for electricity prices, in the wake of the failure of the nation's nuclear power program, is also unfavorable. Many utility companies are faced with very large costs for repairs to their older nuclear stations, while newer stations have encountered huge cost overruns, creating massive local rate increases. Abandoning the nuclear program has led to a renewed use of coal. But the current return to coal-powered generation has forced utilities to confront, at last, the critical problem of acid rain. A solution to acid rain will require fitting both old and new coal-burning plants with expensive equipment, the cost of which will be reflected in higher rates. The prospect of continued rapid increases in the cost of electricity makes the savings in electrical consumption of the ESSPS house especially important.

It is generally believed that oil prices will continue to decline somewhat during the late 1980s. But the present lull in the rise of the price of oil has been created by the dramatic price increases of the past dozen years. These increases have led to the intensive exploration for and development of oil sources located in remote and inaccessible regions, including the North Sea, Alaska, and the Canadian Arctic. Fewer opportunities for discovery of low-cost oil remain than existed before oil prices began to rise in the 1970s, and once a fresh surge of demand develops in an expanding world economy, prices will increase to bring supply and demand back into balance.

The forces that will control the future of energy prices remain as unpredictable as those that created the price changes of recent years, but the underlying conditions of energy supply make continued price escalation certain. Even at relatively low rates of inflation the ESSPS house will produce substantial annual savings for the homeowner (Appendix D). If energy prices increase at rates similar to those experienced in recent years, extremely large savings will be realized.

PART V

SITE DESIGN AND SUBDIVISION PLANNING

15

Site Planning

Entrance and Garden Spaces

Site and landscape design are integral parts of the architecture of the partially earth sheltered house. The exterior spaces that surround the house are closely linked to it by the earth forms that shape the building outline. These spaces can be planned in a great many different ways. Emphasis can be placed on enclosing the exterior space, with entrance courtyard and garden defined by high, encircling berms. The site can also be kept relatively open and the earth berms or banks that protect the house walls tapered away from the building so that they define the space around the house to a limited extent. The design of the entrance courtyard and garden may also emphasize the use of planting screens and plant masses rather than earth forms. Plants can be used instead of berms to outline one or more sides of the enclosures that define the entrance and garden. Still another alternative is to use plantings and earth forms together by placing the planting screens on the top and sides of low berms.

Grade conditions will frequently determine the forms of the enclosing spaces. On steeply sloping ground, embankments cut as the site is leveled for the house and parking area will define the approach to the building. On flatter ground, the entrance may be left open, or it can be defined by enclosing earth forms and plantings. If the entrance courtyard is tightly enclosed, the ground between the enclosing berms and planting screens and the street may be left open to become a front lawn that can be planted with shade trees or a variety of smaller ornamental trees. This space may also be occupied by another berm, or planting screen, creating an intricate, winding approach to the house. Still another option is to enlarge the entrance courtyard, including within it a large area that can become the front lawn or a garden with ornamental plants and trees.

The design of the garden at the back of the house offers the same variety of options. The earth sheltered house does not require the addition of large terraces in order to extend its space out into the garden through its south-facing glass; earth shapes and plantings will perform this function more subtly and at less expense. The patio or deck is placed a short distance away from the house to limit ground reflection, and here it will be in the center of the space defined by the enclosing berms or plantings. If the garden is completely enclosed by high berms, it becomes a sheltered space that is comfortable in moderately cool weather; the berms will also shelter the exposed south wall of the building from wind during the coldest months of the year. Lower enclosures allow views to open beyond the partially enclosed garden, giving greater depth to the lot than can be achieved either by tightly enclosing the garden, or by opening the backyard completely. High and low enclosures can also be combined; an intimate, protected area on one side of the lot will shelter the patio, while the other end of the garden opens out to the view.

On a small suburban lot, the enclosing berm or plantings that define the garden will usually extend to the rear property line. If the lot is larger, multiple spaces can be outlined with earth forms and plantings, and this will, again, give greater depth to the lot than an open design. Where the back of the house set on a larger lot is kept open, with berms tapered sharply into the natural grade close to the house, it will usually be desirable to provide enclosure along the rear property line (figure 15–3). Enclosure at the rear can be achieved with a berm or by establishing a planting screen. A berm 3 to 4 feet high built across the back of the lot will provide a substantial amount of privacy, which will be increased when plants established on top of the berm mature.

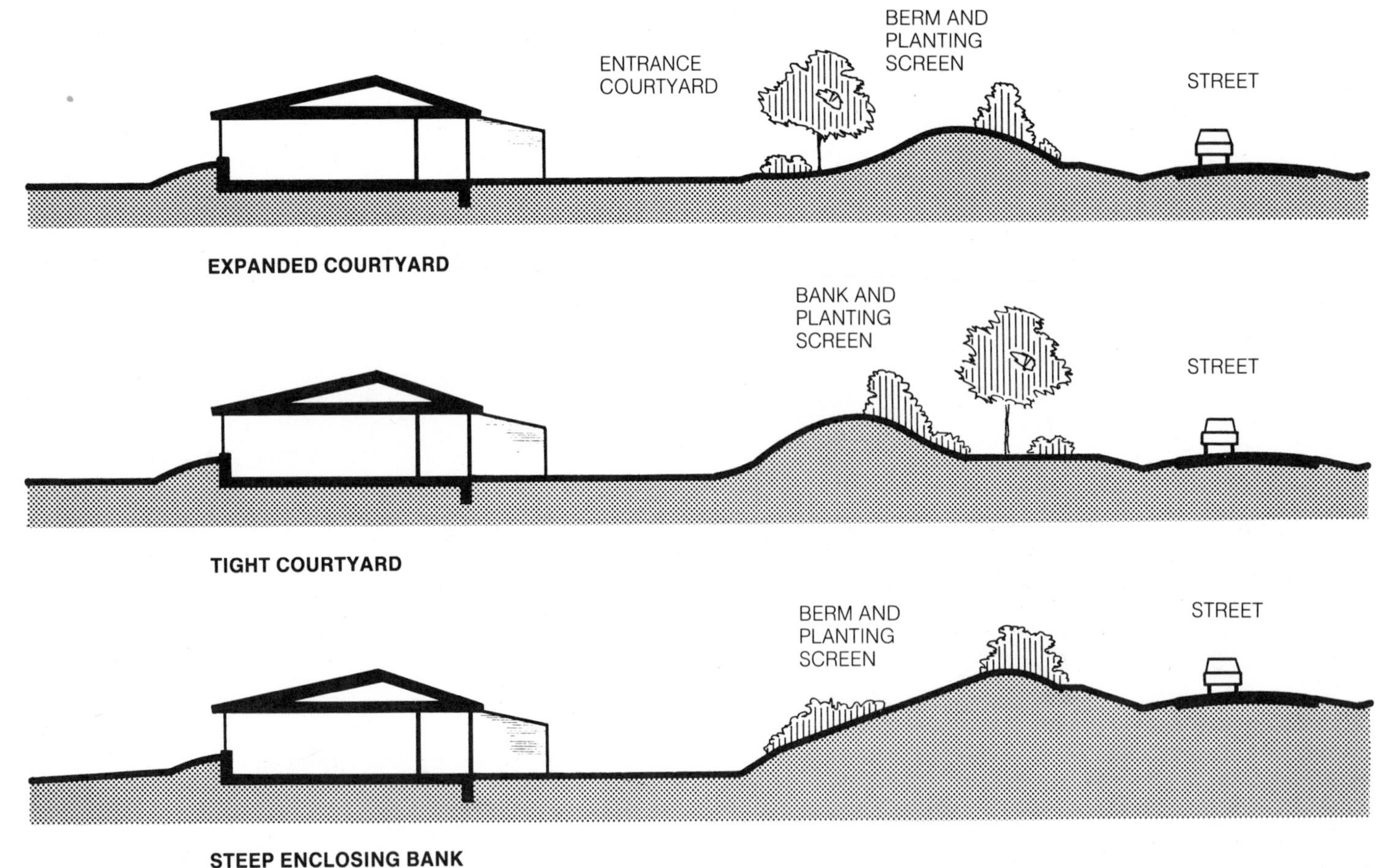

15–1. Enclosing entrance spaces (sections).

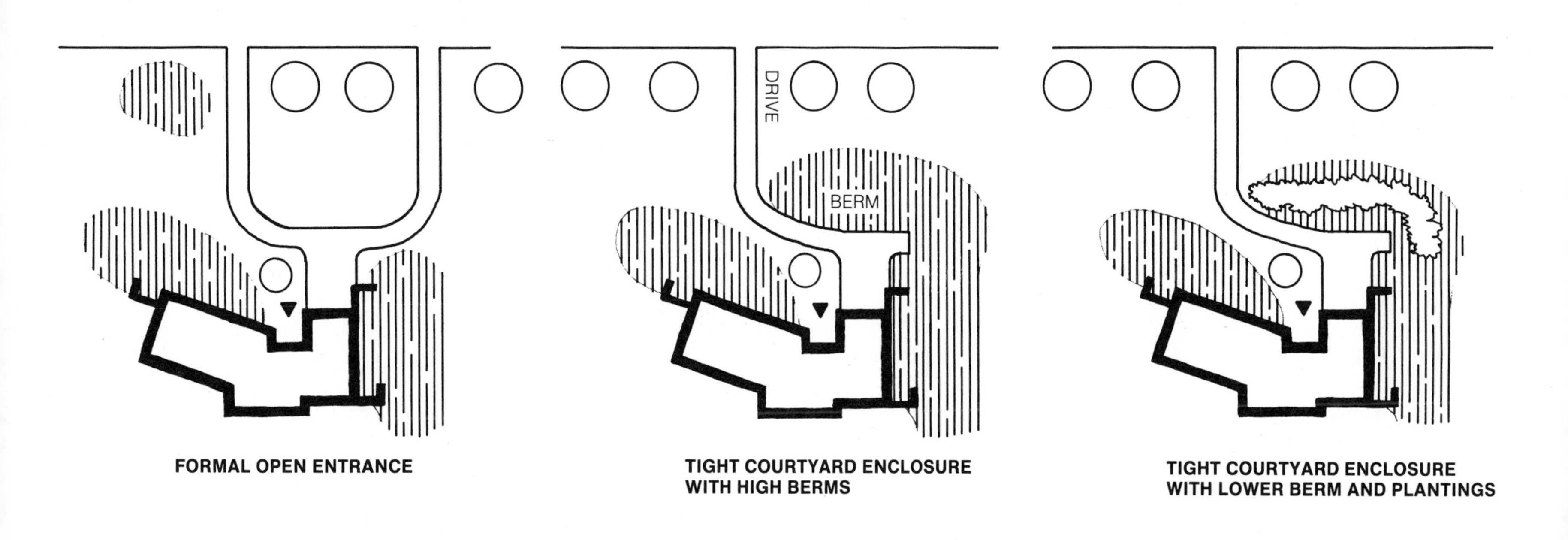

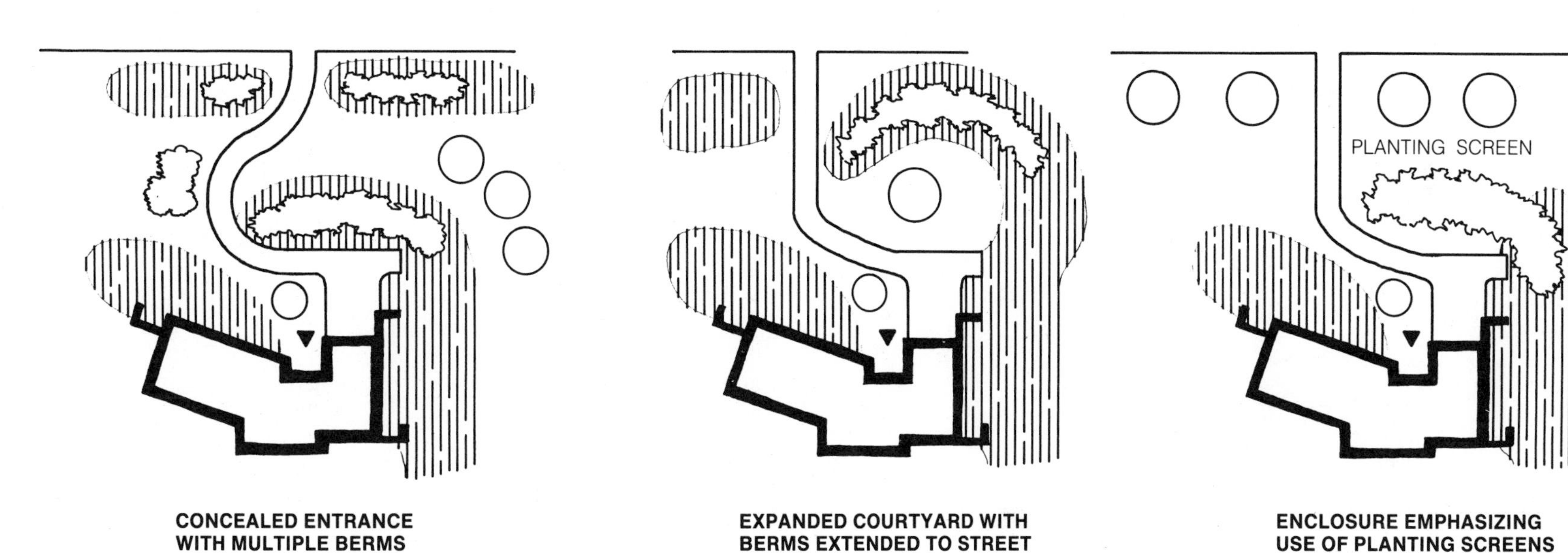

15–2. Enclosing entrance spaces (plans).

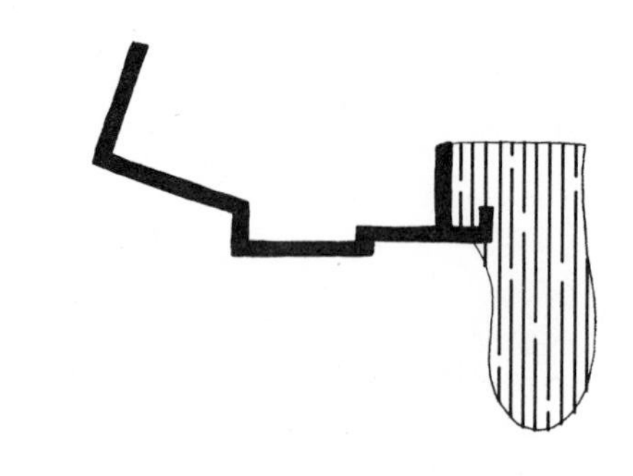

OPEN GARDEN

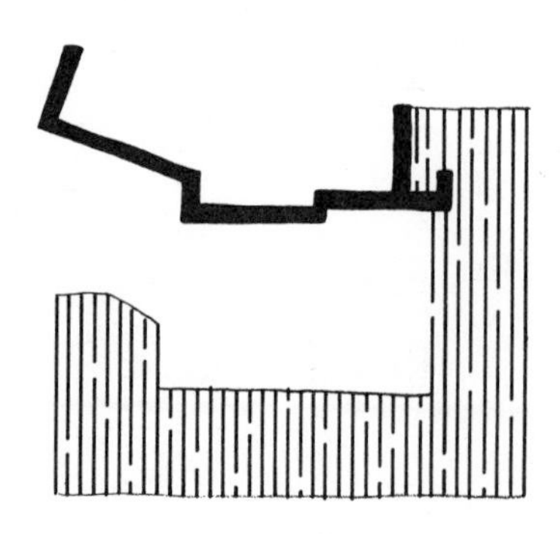

ENCLOSURE WITH GEOMETRIC BERM

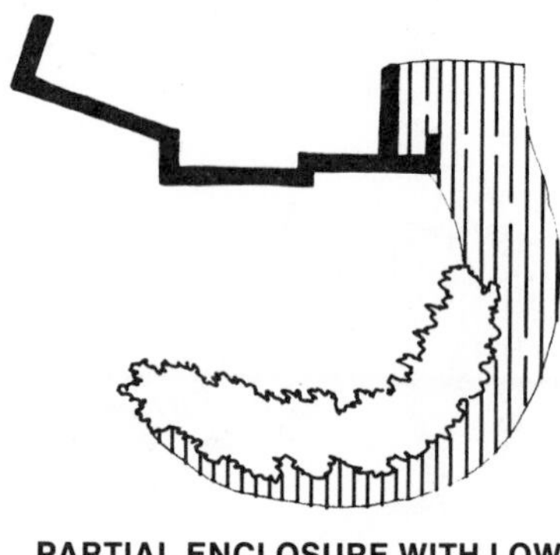

PARTIAL ENCLOSURE WITH LOW BERM AND PLANTINGS

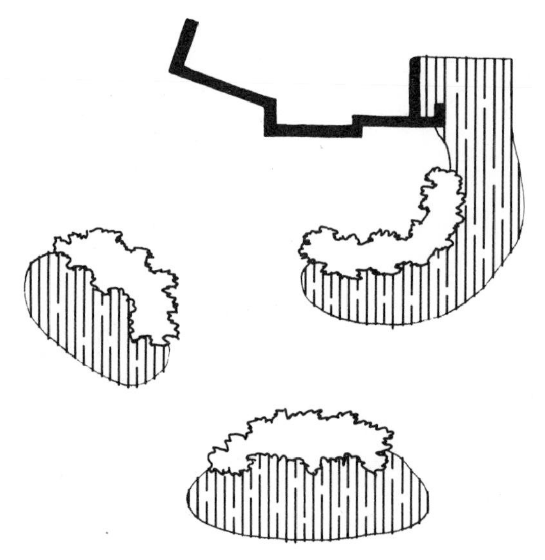

WINDING GARDEN SPACE ON A LARGE LOT

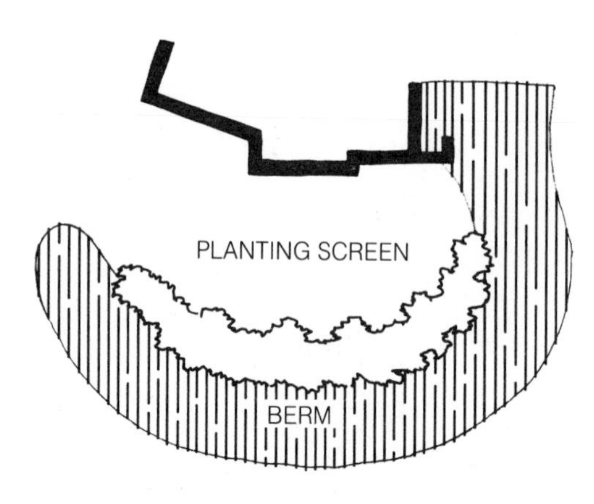

FULL ENCLOSURE WITH BERM AND PLANTINGS

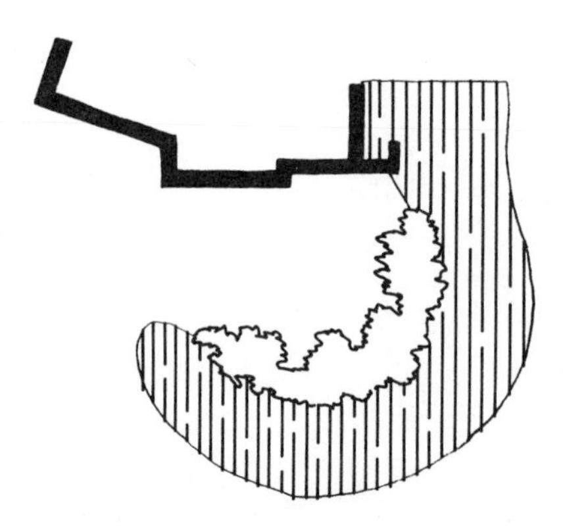

PARTIAL ENCLOSURE WITH HIGH BERM AND PLANTINGS

15–3. Garden enclosures.

Handling Grades—the Flat Site

Siting the ESSPS house requires careful attention to grade conditions. The building must be designed so that its major glass areas face south. The house must also be closely fitted to the slope of the land, in order to avoid extra foundation work and excessively high retaining walls or berms. Site planning will be easiest on a south-facing slope, where the house fits comfortably into the hillside. Sites of all types can, however, be used, and the ESSPS house is readily adapted to east- and west-facing hillsides, to north-facing slopes, and to flat ground.

Where the partially earth sheltered house is built on a flat, or nearly flat site, there is a great deal of freedom in the design of the earth forms that surround the house. Steeply sloping berms can be used to make the building stand out sharply from its site (figure 15–4). This gives the building a formal, geometric quality, and formal landscaping will be needed, both on the slopes of the berms and in the areas surrounding the house. A subtler handling of the ESSPS design, which emphasizes the close relationship of land and building, can be achieved with berms that are graded away from the house more gradually and are designed to blend into the surrounding site. These berms will appear to rise naturally out of the land if the entire lot is regraded with slightly rolling contours.

If an earth sheltered house is built by itself, in a conventional subdivision, berms

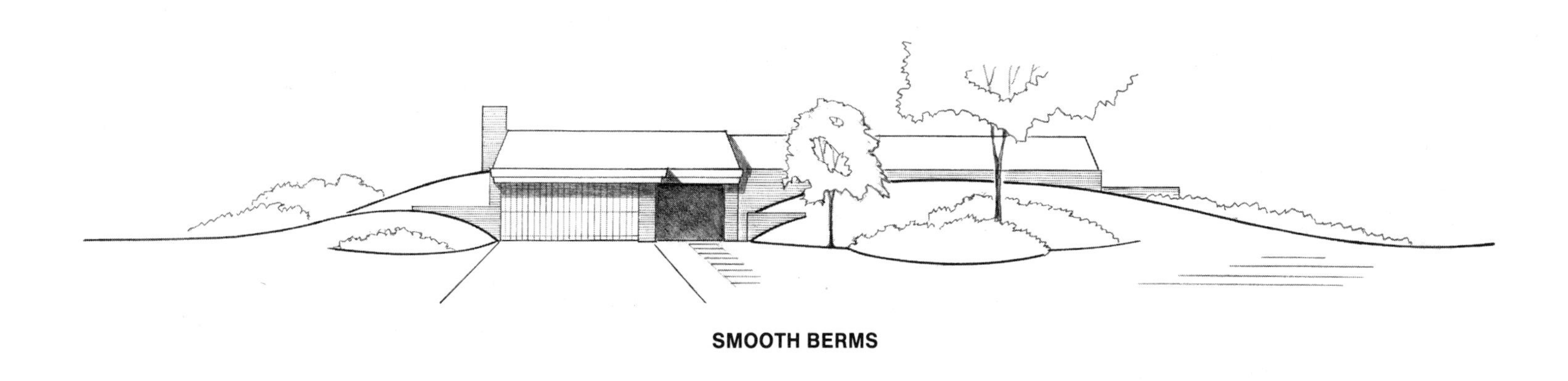

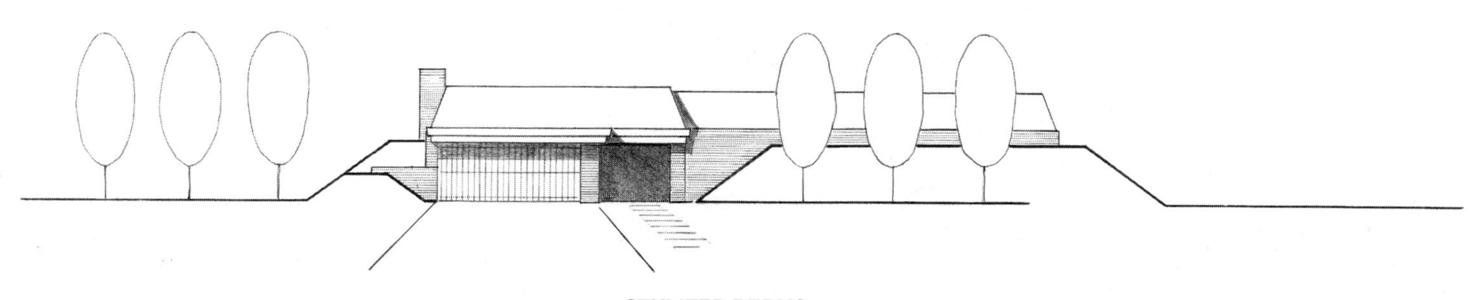

15–4. Smooth and stylized berms.

extending a distance from the house to enclose the garden and entrance courtyard should be lower than those built up against the house, so they do not become arbitrary landscape features. Regrading at a distance from the house should be just sufficient to give the land a three-dimensional quality, allowing the high berms that protect the building to become natural extensions of the contours of the site (figure 15–5). This will usually make it difficult to develop enclosed exterior spaces using earth forms alone. To enclose the garden and courtyard on a flat site, low berms should be combined with planting screens. Partially earth sheltered houses built as a group can, however, have higher berms enclosing the gardens and courtyards that extend along the street, since these berms will establish the character of the regraded terrain.

The transformation of a flat lot into contoured terrain must be planned so that the amount of earth obtained from cuts balances the amount used to form the berms that shelter the house and give the lot its three-dimensional quality. If this is not done, grading will be very expensive.

A half-acre lot can be completely regraded, with an average change in grade of 5 or 6 feet, for only a few hundred dollars. But hauling in even a small portion of the earth moved in a regrading operation of this type can cost several thousand dollars—an unnecessary expense that will be completely outside the limits of the grading and excavating budget.

Balancing cuts and fills on a flat site requires grades to be lowered on part of the lot, in order to provide the fill for the berms. The entrance area to the house, including the garage, courtyard, and parking area, ordinarily should not be lowered, since this can create drainage problems for the garage and parking apron. The back of the house provides a better source of fill. The area occupied by the garden can be cut down 2 or 3 feet from the natural grade. If low berms are built around the excavated area and the rest of the fill is used to protect the house walls, an enclosed sunk garden will be formed with a minimum of excavating work. This design will create a change in level between the house and garden. Alternatively, the house slab can be lowered to the level of the garden. Here the change in level occurs between the garage and entrance and the house proper, rather than between house and garden. Lowering the slab elevation will provide additional fill, while reducing the amount of earth needed to form the sheltering berms.

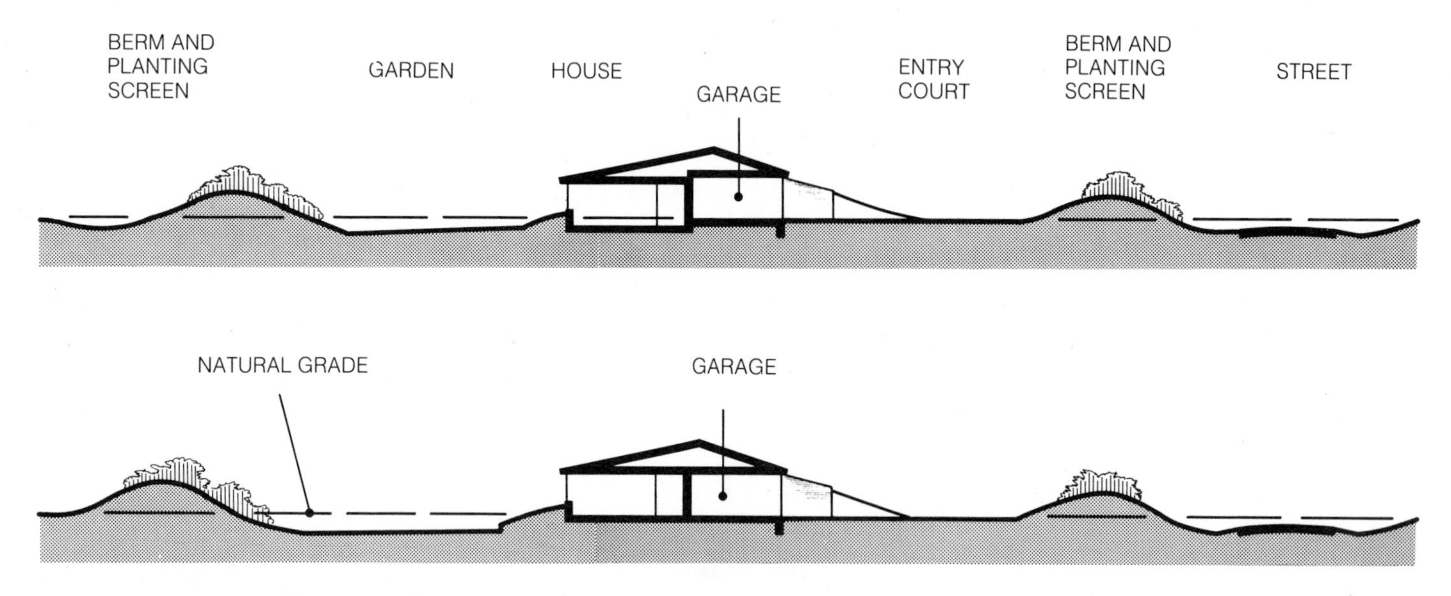

15–5. Using berms to shape flat terrain.

South-facing Slopes

The south-facing slope presents few problems as long as steep grades are not encountered. On a gradual south-facing slope, a wide courtyard can be cut into the hillside (figure 15–6). Fill from the enclosed entrance area can be used for berms that protect the walls of the house and for leveling the garden on the downhill side of the house. Steeper south-facing slopes can be much more difficult. The house will continue to fit easily into the hillside, and if the grade is steep enough, the north wall will be entirely protected by the natural grade. A north-side entrance courtyard, however, will require deep cuts. Very high, steep banks are formed on the uphill side of the courtyard, and large amounts of fill will be removed as the courtyard is cut out; this material must be placed on the low side of the site, where it will help level up the garden. The steep slopes of the courtyard embankment and the mounded garden require careful landscaping to prevent runoff and erosion problems (figure 15–7).

These problems can be avoided if the entrance area is redesigned. Placing the entrance at the end of the house, rather than on the north side, will help avoid deep cuts around the courtyard, but a steep driveway, leading down from the street above, will frequently be required. The driveway problem can be solved by creating a change in level betwe›n the entrance area and the rest of the house. On some sites, the change in level may be only a few steps, but where the house is built on a very steep hillside the entrance can be set a full floor above the rest of the house. If the entrance courtyard is designed in this manner, the entire building, except for the garage and upper-level foyer,

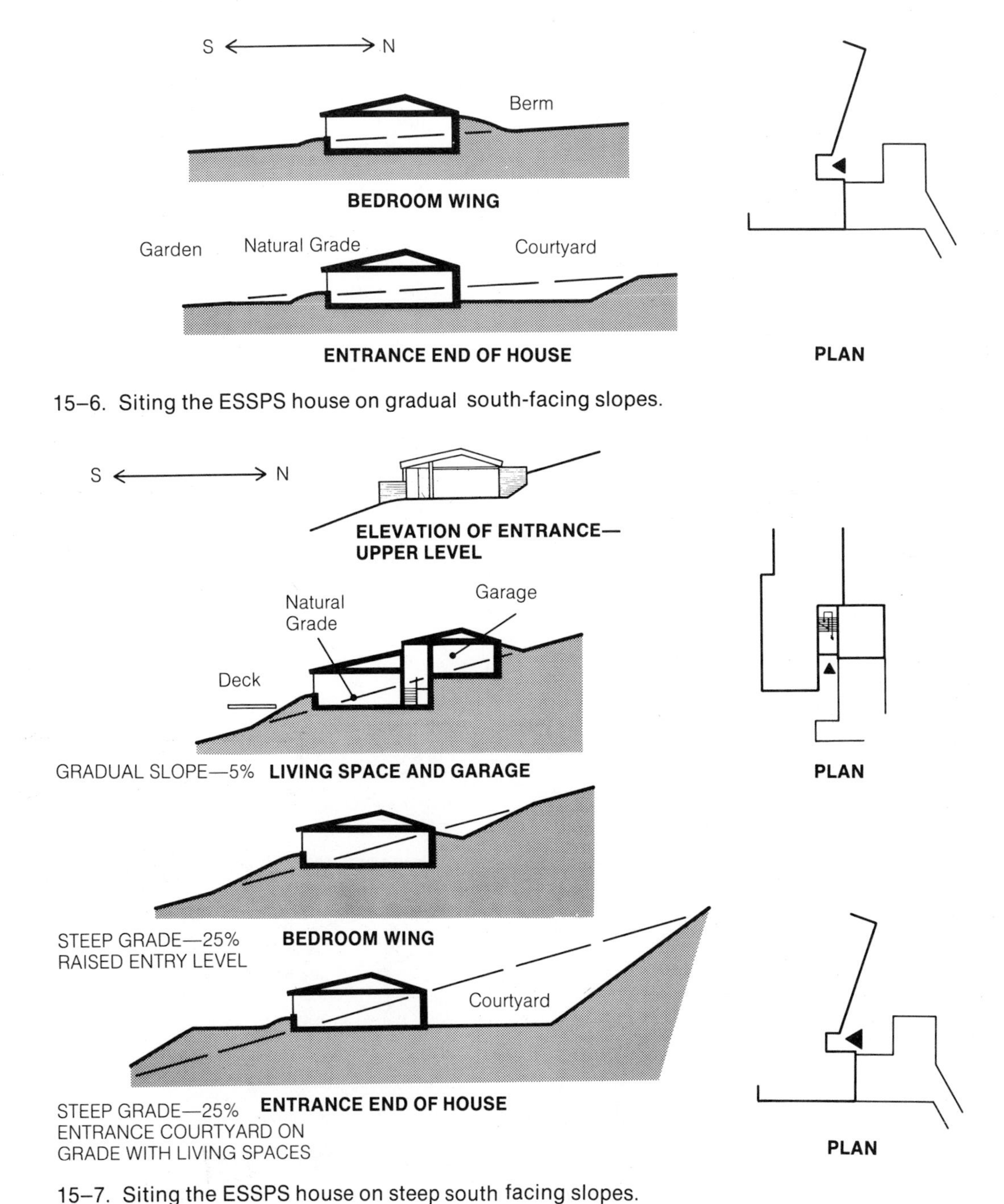

15–6. Siting the ESSPS house on gradual south-facing slopes.

15–7. Siting the ESSPS house on steep south facing slopes.

will be concealed on the approach side. The house will only be discovered by following the flight of steps that leads down from the parking level to the living areas and garden. South-facing grades as steep as 20 percent to 25 percent can be used if a raised entrance court of this type is incorporated into the design.

East- and West-facing Slopes

On hillsides that face east and west, the length of the house, rather than its width, runs counter to the fall of the land. One end of the house now fits into the hill; the other end is exposed and must be protected by an artificial berm. Excavation is required both for the entrance courtyard and to level off the garden, and ample fill is available for berms.

Houses built on east- and west-facing slopes can have their entrances at either the uphill or downhill end of the house. If the garage and parking areas are at the uphill end, a courtyard is defined by natural embankments. Fill from the courtyard protects the closed north wall of the house, which will have relatively little natural earth cover. If the lower side of the lot is chosen for the entrance area, the courtyard has low embankments and can be either open or closed. With either arrangement, sites that have grades of 10 percent can be used without introducing changes in level. On steeper sites, the house can be fitted to the slope of the land by changing the level between the living spaces and the bedroom wing. This will allow east/west grades in excess of 15 percent to be handled without difficulty (figure 15–8).

North-facing Slopes

The north-facing slope presents the most serious challenge for the designer of the ESSPS house. Gradual north-facing slopes, with grades that do not exceed 5 percent or 6 percent, are not difficult, but when the north slope approaches a grade of 10 percent, the house must be carefully fitted to its site; steeper north-facing hillsides should generally be regarded as unsuited for the ESSPS house.

On gradual north slopes, the entrance area can be placed at the end of the house, or on the north (low) side of the building (figure 15–9). Fill taken from the garden is used to level up the courtyard, and low berms can be constructed partially to enclose the entrance. Enclosing the courtyard more completely with earth forms is difficult; retaining

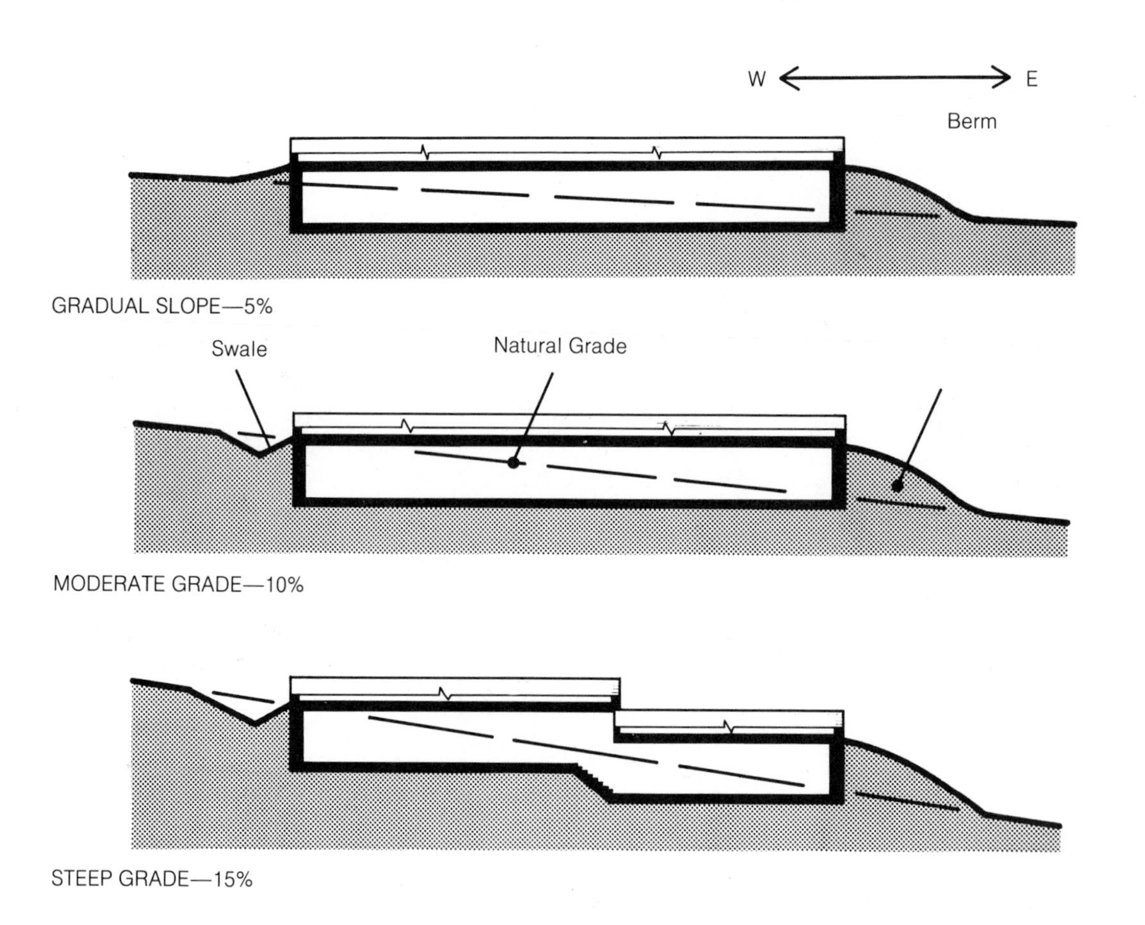

15–8. Siting on east- or west-facing slopes.

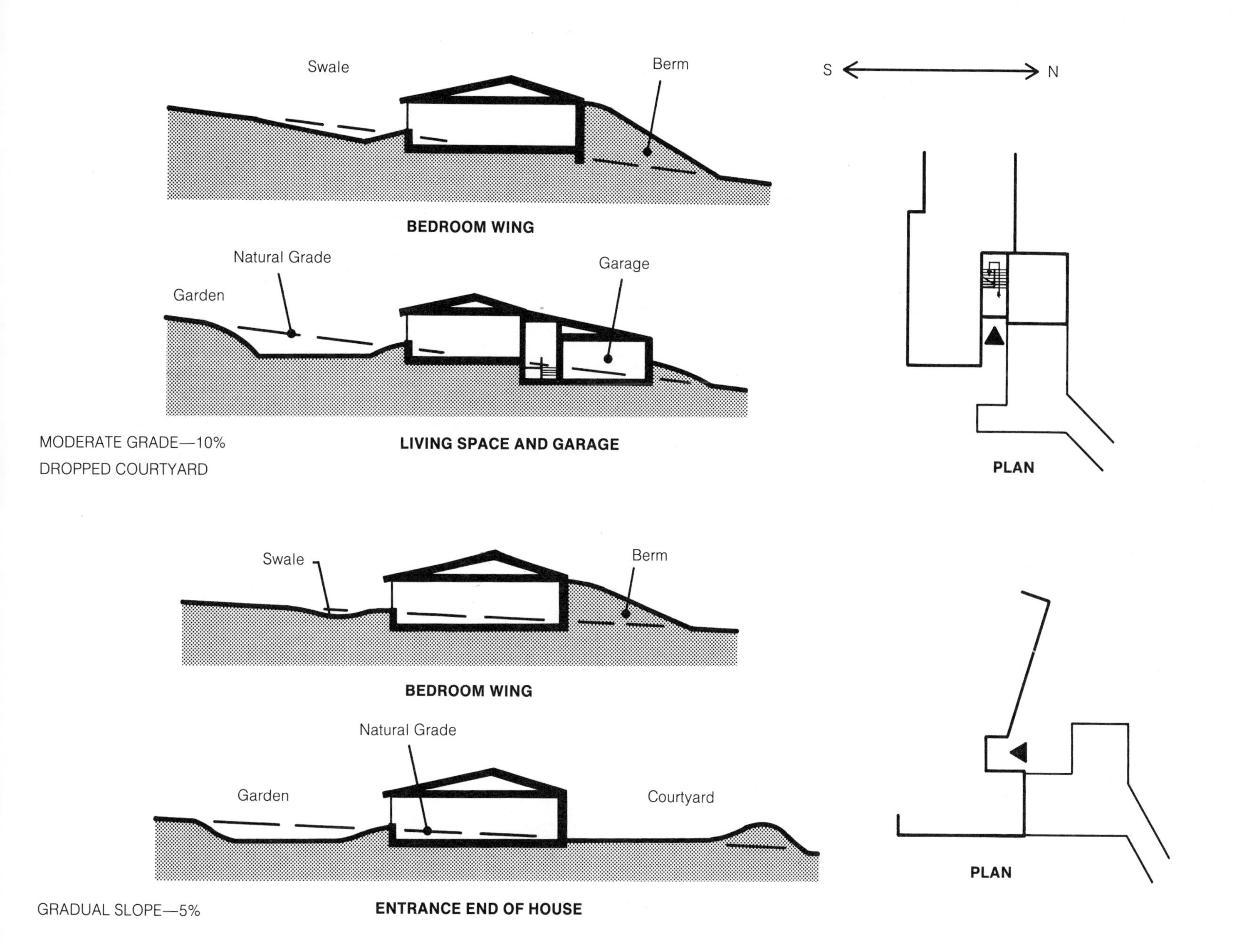

15–9. Siting the ESSPS house on a north-facing slope.

walls or planting screens will be more practical than berms. The garden must be cut out of the slope on the uphill side, and it will usually be fully enclosed by natural embankments, so that a drain line is needed to carry runoff water under the house to the low side of the building.

On steeper north-facing slopes, slab elevation must be kept up, to avoid sinking the south wall and the garden too deeply into the hillside. The north wall of the building requires a foundation that extends below the slab, and the protective berm becomes a very high berm. This high and somewhat unsightly berm can be masked by an end entrance that places the garage between the entrance court and the berm, although the berm will still remain visible from neighboring houses to the north. Where the grade approaches 10 percent, the entrance courtyard and garage should be kept below the level of the living spaces, so that they are not built up on fill.

16

Landscaping the Lot

Trees and Winter Sunlight

Trees, more than any other landscape or architectural element, determine the appearance of the residential street and garden. Trees also have a significant environmental impact. During the nineteenth century, the streets of the new communities established as the nation grew westward were planted with rows of lofty shade trees. Later, the streets of many suburban communities were also lined with oaks, maples, and elms. Today the tree-lined street is the most familiar image of the American small town and suburb.

The establishment of rows of shade trees ended in many suburban areas with the introduction of central air-conditioning. Recently, however, there has been renewed interest in the use of trees to modify the building's environment. Windscreens on the north side of the house have been suggested, and there has been a new emphasis on using shade trees to limit the need for increasingly costly mechanical cooling.

The ESSPS house, with its protective earth banks on the north side of the house, airtight envelope, and summer sunscreens shielding the windows, will obtain only minimal energy savings from protective tree plantings. For the ESSPS house, trees are instead a possible source of energy loss, since shadows cast in winter by improperly placed trees will reduce passive solar gain. If all interference with passive gain is to be avoided, trees must be kept clear of the building, within the arc described by the late winter or early spring sun—the sunlight arc in late fall and early winter is much shorter (figure 16–1). This will keep large trees at a considerable distance from the south side of the house. At the winter solstice, the noon sun at 40 degrees north latitude is only 26 degrees above the horizon. In this latitude, trees that reach 40 feet at maturity must be planted over 90 feet from the building to avoid casting shadows on the south wall; taller trees must be planted at a greater distance. Morning and afternoon sun height in winter is much lower, requiring trees planted to the southwest and southeast of the house to be located even farther from the building, so their shadows will not fall on its windows.

Tree shadows falling on a portion of the glass area of the ESSPS house during the course of the day will, however, have little effect on the overall performance of its passive solar system. This makes it possible to plant deciduous trees in properly chosen locations within the winter sunlight arc. The size of the shadow area cast by bare tree branches will vary with different tree species, as well as with the size and age of the individual tree. Shadow area may, in some instances, be as high as 50 percent of the area shaded by the tree when it is in leaf, but frequently it will be much smaller than this. An overall estimate of one-third of the shade area, for winter shadows cast by commonly planted tree species, seems reasonable. Tree branches will only block direct radiation; indirect radiation will not be affected unless the tree is very close to the house. Table 16–1 gives the effect on passive gain of winter tree shadow, assuming that enough trees are planted to cast shadows on the entire glass area of the house throughout the winter daylight hours.

Shading all the glass area of the house for the entire day will reduce net gain on a clear day by over 20 percent in a relatively mild climate, with an even greater loss in a more severe climate.

The impact on the performance of the passive solar system of the ESSPS house will, however, be relatively small, even where, as in Table 16–1, all of the glass area is affected by tree shadows. Maximizing clear-day gain is of little significance for the ESSPS house, since on clear days far more energy is acquired than can be used by the storage system, and the energy lost because of tree shadows would in any case be dumped. For the ESSPS house, the critical days are the partially cloudy or very cloudy days when passive gain drops below the level required to meet daily heat loss. When these weather conditions prevail, the propor-

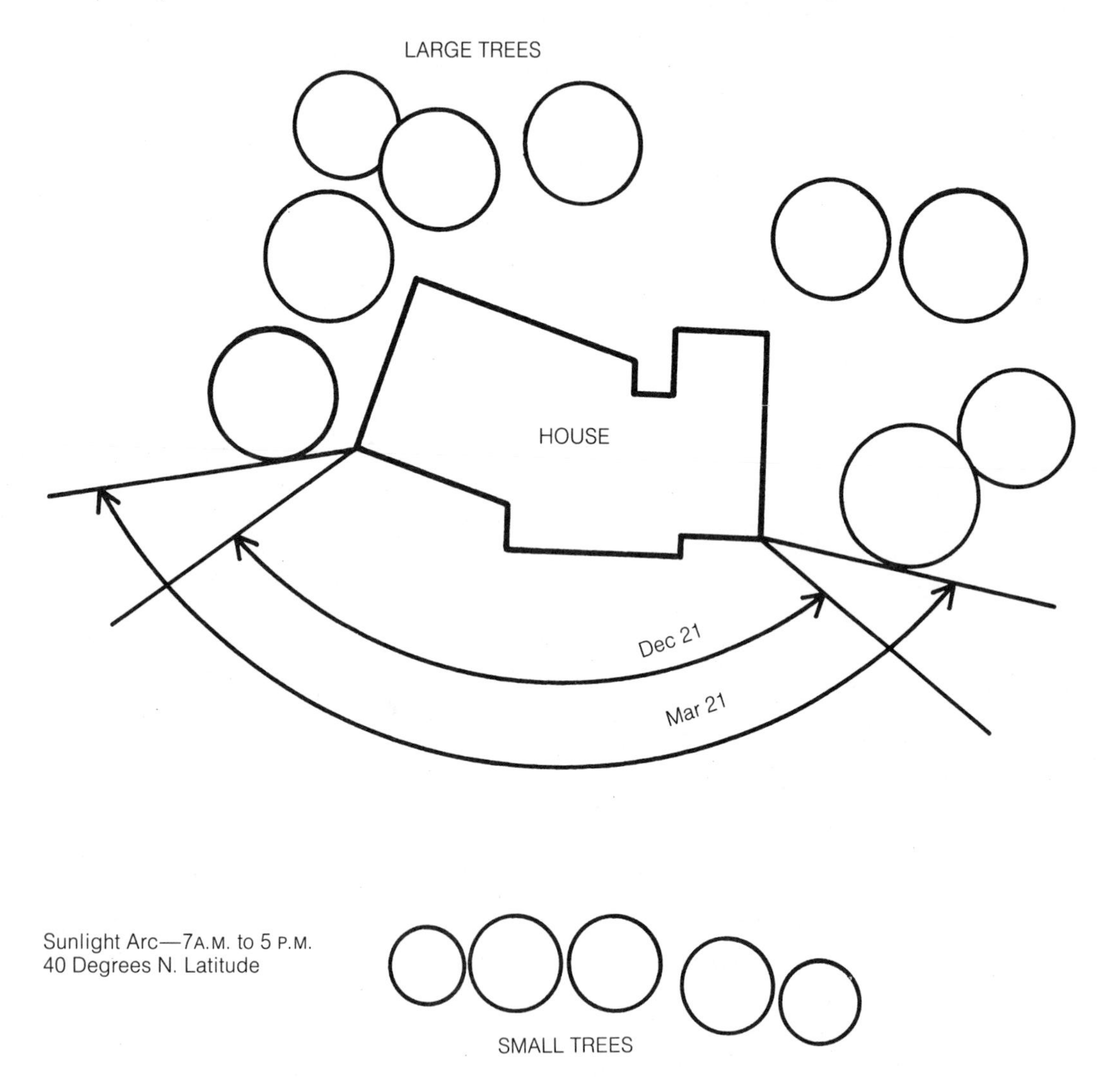

16–1. Winter sunlight and the location of trees.

Table 16–1. Effect of Tree Shadows on Clear-day Winter Passive Gain

	Unobstructed Glass	Glass with Winter Shadows
Direct radiation*	1,103	739
Indirect radiation*	543	543
Total solar radiation	1,646	1,282
Transmission loss (20%)**	329	266
Heat loss from house (10 hours)**	176	176
Net daily gain	1,141	840

* Solar gain for Dec. 21, 40 N. latitude in Btu/sq. ft. of glass.
** Average outside air temperature 34° F.

tion of indirect to direct radiation climbs sharply. On very cloudy days almost all of the passive gain may come from indirect radiation, which will not be affected by the presence of trees, unless they are very close to the building.

Limited intrusions into the winter sunlight arc are therefore permissible. A few large shade trees can, for example, be planted close to the southern property line. If the trees are located in a corner of the lot, they will only shade the south wall for half the day, and their shadows will cover only one-half or less of the glass when they do fall on the house. Small ornamental trees, which are usually thin branched, can be placed closer to the house, in the garden. Their thin branches will block little winter sun and will form welcome pools of shade in summer. Larger trees can also be planted at the ends of the house. Here, they will have a minimal effect on winter passive gain, since they will only interrupt weak early morning and late-afternoon sunlight; in summer their foliage will effectively shade the glass on the east and west walls of the house. In addition to this selective planting within the winter sunlight arc, trees can be placed around the rest of the house without concern for their impact on passive gain.

Garden Design

The landscape design includes, in addition to trees, lawns, ground cover, and plants that are used both for ornamental effect and for planting screens. Ground cover should be used as extensively as the budget permits, but where berms must be kept largely in grass they should be shaped to permit safe and easy use of a power mower. With a limited budget ground cover may be restricted to the front entrance area and the berms under the windows on the south wall; if the budget is somewhat more generous, ground cover can be planted along the entire length of the protective banks or berms on the north side of the house, as well as on berms that define the courtyard and garden.

Berms are also the natural location for planting bushes and ornamental trees. Where berms are planted to establish screens that define exterior spaces around the building, plant materials must be selected with care. A solid, all-year planting screen should be made up primarily of evergreen trees and bushes. Evergreens grow slowly, but screens planted on low berms will usually need to be only 3 to 4 feet high, and most species will reach this height a few years after they are planted. The plants, however, must be placed only a few feet apart if they are to fill in quickly to form a solid screen. Good quality evergreens, closely spaced, will absorb a large share of the landscape budget if screens of any length are required. One solution is to establish more rapidly growing and less expensive screens with inexpensive deciduous plants, such as the familiar fire thorn, or pyracantha, or the privet hedge. Generously spaced evergreens can be placed in front of these deciduous plants; once they have matured they will supplement or replace the deciduous screen, which, at least in winter, will be less solid and less attractive than the evergreens.

Extra emphasis should be placed on landscaping the visual focal point at the entrance to the house. An entrance planting design for a modest budget can be developed around a single ornamental tree of moderate size, such as a delicate clump birch, or an American holly. Flowering bushes, set in a bed covered with wood chips, and ground cover, planted so that it fills in rapidly, will complete the basic landscaping of the entrance.

An ornamental walk, with brick or stone paving, adds greatly to the quality of the entrance design and should be part of even a modest construction budget. The walk can be kept narrow—a large display of paving will add little to the design.

Relatively inexpensive and attractive walks can be built with flagstone laid directly on the ground, although the stones will settle during the first year after construction has been completed and will usually require resetting. Brick walks are more expensive, but have a special charm, which is enhanced over the years as the brick mellows.

The least expensive outdoor sitting area is a deck supported by ordinary framing timbers and covered with pine or fir planking. Where pressure-treated lumber can be substituted, a more durable deck is created at little additional cost. Redwood decks are very handsome, but they cost a great deal more than a deck built of pressure-treated lumber. Patios can be built for little more than a redwood deck, and should be used, wherever possible, for long-term durability. Plain concrete, with redwood dividers, is frequently used to meet the requirements of a limited budget. If properly designed, both walks and patios of this type are very attractive. A great deal will, however, depend upon workmanship, and the final product can often be disappointing. Flagstone and brick are the most appropriate materials for the patio of the earth sheltered house, and if the patio is kept no larger than is needed for actual use, it will usually be possible to find the money for them, if only as a modest extra added to the original budget of the house.

17

Subdivision-planning Methods

Finding a Lot

Finding a site for the ESSPS house in a conventional subdivision will often be very difficult. If the house is built on country acreage, or on an exceptionally large suburban lot occupying several acres, the problems of building orientation and tree shadows can usually be solved and the characteristics of the ESSPS house do not restrict the choice of ground. Subdivision lots of more usual size (¼ to 1 acre) will be more difficult to use; most will not be suitable for the ESSPS house.

In a conventional subdivision, the lot chosen for the ESSPS house should be located

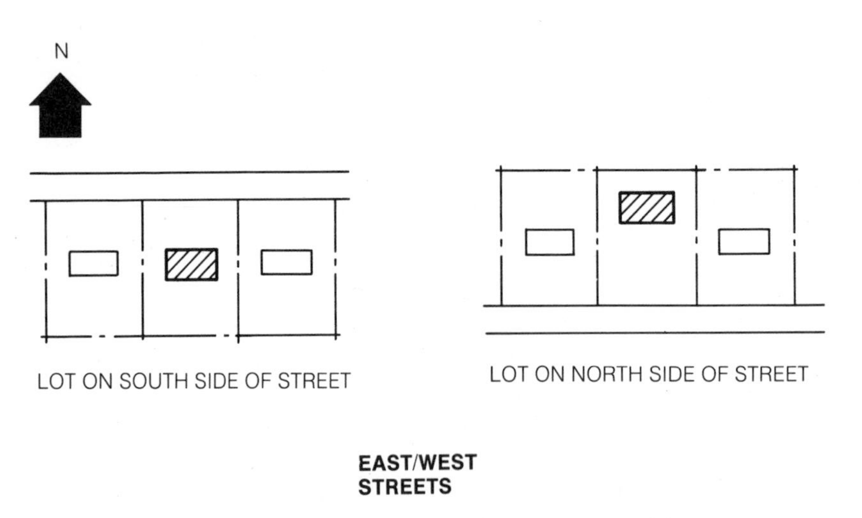

17–1. Placing the ESSPS house in a conventional subdivision.

on the south side of a street that runs approximately east and west, so that the open south side of the building will face the rear of the lot (figure 17–1). If a lot on the north side of the street is selected, the house must be placed as far to the rear of the lot as zoning restrictions permit, and, unless the property is large, the garden will be hemmed in by neighboring homes built closer to the street. North/south streets in the subdivision must be avoided, since here the ESSPS house will be turned sideways on the lot, and it will face directly toward its neighbor on the south. These considerations will eliminate at least three-fourths of the lots in the conventional tract.

Lot frontage must also be wider than usual, and, in order to obtain the needed lot width, it will often be necessary to purchase a larger plot than is desired. Steep grades, especially on north-facing slopes, will usually be difficult, and wooded ground should not be selected. Extensive reshaping of the lot is required with earth sheltered construction, and most of the trees on the property will be destroyed when the lot is graded. In an established subdivision, mature trees located on neighboring properties may block large amounts of winter sunlight, and in a new subdivision trees planted by the neighboring homeowners may also eventually block a large part of the winter clear sky area.

Finding a lot for the ESSPS house in a conventional subdivision will therefore require an extended and sometimes unsuccessful search. If ESSPS houses, with their passive solar systems and partial earth shelter, are to be built in large numbers, it will be necessary to develop subdivisions that are planned to meet the requirements of this new type of building.

Suburban-planning Concepts

Subdivisions intended for the ESSPS house will be very different in character from the conventional subdivision. The layout of their streets and lots must be designed to provide all the lots with properly oriented building sites, adequate frontage, and relatively unobstructed winter sky. Perhaps more important the appearance of the completed community will be radically altered. Houses will be partially or wholly hidden from the street, and landscape views will replace the long rows of evenly spaced one- or two-story building facades that line the streets of the conventional subdivision.

The suburban street, as it exists today, is the product of an idealized image of the American small town or city of the mid-nineteenth century. Rows of houses set behind lawns are built facing the street, which until recently was heavily planted with shade trees. As long as the suburbs remained small, the suburban community, like the small town upon which it was modeled, often had a great deal of charm. But in the last forty years, the suburbs of the American city have grown to enormous sizes, and the regions surrounding cities, formerly occupied by farms, woods, country estates, and small suburban settlements, have been turned into a vast area of continuous tract development. Within this massive suburban ring, open views and a sense of vista have vanished, replaced by mile after mile of streets lined with regularly spaced houses. Differences in the size of their lots and homes distinguish one subdivision from another, but the dominant impression in suburbia is an endless repetition of local streets and rows of houses.

American traditions of subdivision planning have contributed to the problems of suburban sprawl. Subdivision planning in the United States has paid little attention to the need for the retention of open space within the suburban community. Since the start of suburban development in this country, over one-hundred years ago, imaginative architects and landscape planners have proposed community plans that incorporate park areas, paths, lakes, and other elements of visual variation. These plans sought to preserve the most beautiful parts of the landscape in their natural state and provide community recreation facilities. Planners have also understood the need to preserve large areas of open space within the sprawling suburbs of each metropolitan region in order to separate local communities and provide them with visual definition as the suburbs expand. These ideas, however, have remained in the realm of theory, taught in schools of planning and architecture, but having very limited influence on development practice. With only rare exceptions, the design of subdivisions for single-family houses has been restricted to developing a simple layout that creates uniform lots covering the entire tract, without open space, or other elements of visual variation.

Earth sheltered/passive solar design offers the opportunity to change the character of the suburban street within the confines of existing subdivision development practice. Developers of tracts designed for the ESSPS house may be no more inclined to adopt imaginative planning methods than are the developers of conventional tracts. But even with a simple division of ground into lots of uniform size, the earth sheltered suburban community will retain a strong sense of its original landscape and present continuing variety in its local streetscapes. Architects and builders will no longer be compelled to achieve variety on the suburban street by placing a bit of brick on one house and stone on another. Instead, there will be a much wider range of design possibilities derived from the earth forms and plantings that replace repetitive facades as the dominant visual elements in the community.

Designing the Subdivision for New Requirements

The conventional subdivision plan does not pay attention to building orientation (figure 17–2). Street layout is determined by the tract boundaries, and the tract planner has to obtain the maximum number of lots permitted by zoning regulations, while minimizing the amount of road construction within the subdivision. Houses face each other across the streets of the development, with each house sited parallel to the street. If the street runs generally east and west, the house on the south side of the street will have its private living area—the backyard—opening toward the south, but the house across the street will have its backyard on the north side of the building. If streets run generally north and south, one set of backyards will face the east, and the other west.

This varied, or random, orientation will be replaced, in the subdivision designed to use passive solar energy, by a layout that permits all of the homes to face south, with largely unobstructed winter sky, and a private garden or sitting area. Economical tract design still requires fitting the street layout to the tract confines, and streets may run north and south as well as east and west, or have an irregular and informal layout. The relationship of the house and the street will thus change as street direction changes. In the conventional subdivision, every house faces the street behind its front lawn, with a private outdoor area at the rear of the lot. In the energy-efficient subdivision, this arrangement cannot be retained. The backyard will frequently become a front yard, facing the street, or it will turn into a side yard, facing a neighboring house. These varied layouts are possible with earth sheltered architecture, because extensions of the earth forms surrounding the buildings can be used to create private, sheltered outdoor spaces on any part of the lot.

17–2. Typical plan of conventional subdivision.

Orientation—Tracts with East/West Streets

If the energy-efficient subdivision is laid out with streets that run generally east and west, the houses on the south side of these streets can be sited in a conventional manner (figure 17–3). They will have front entrances on their north sides, facing the street, and gardens at the back of the lot (figures 17–4, 17–5). The houses across the street, on the north side must, however, be moved from their usual position and placed at the rear of the lot. Their outdoor living areas are between the street and the house, with the driveway located close to one of the side lines of the lot. The driveway leads to a garage placed behind the house, where it does not block sunlight; the entrance will be on the side of the house.

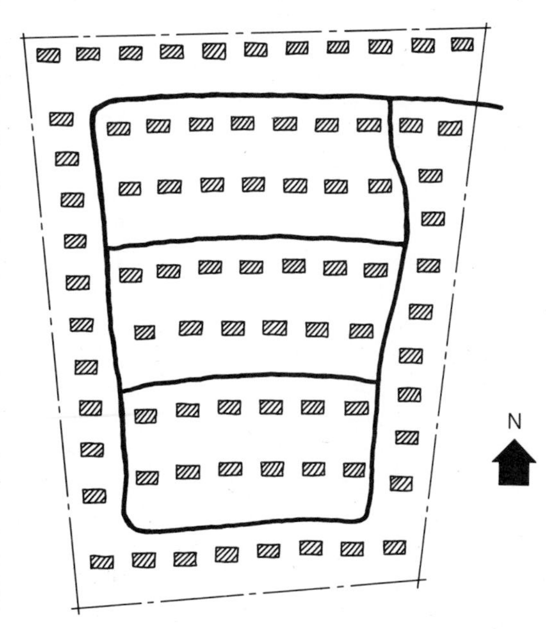

17–3. ESSPS subdivision plan with east/west streets.

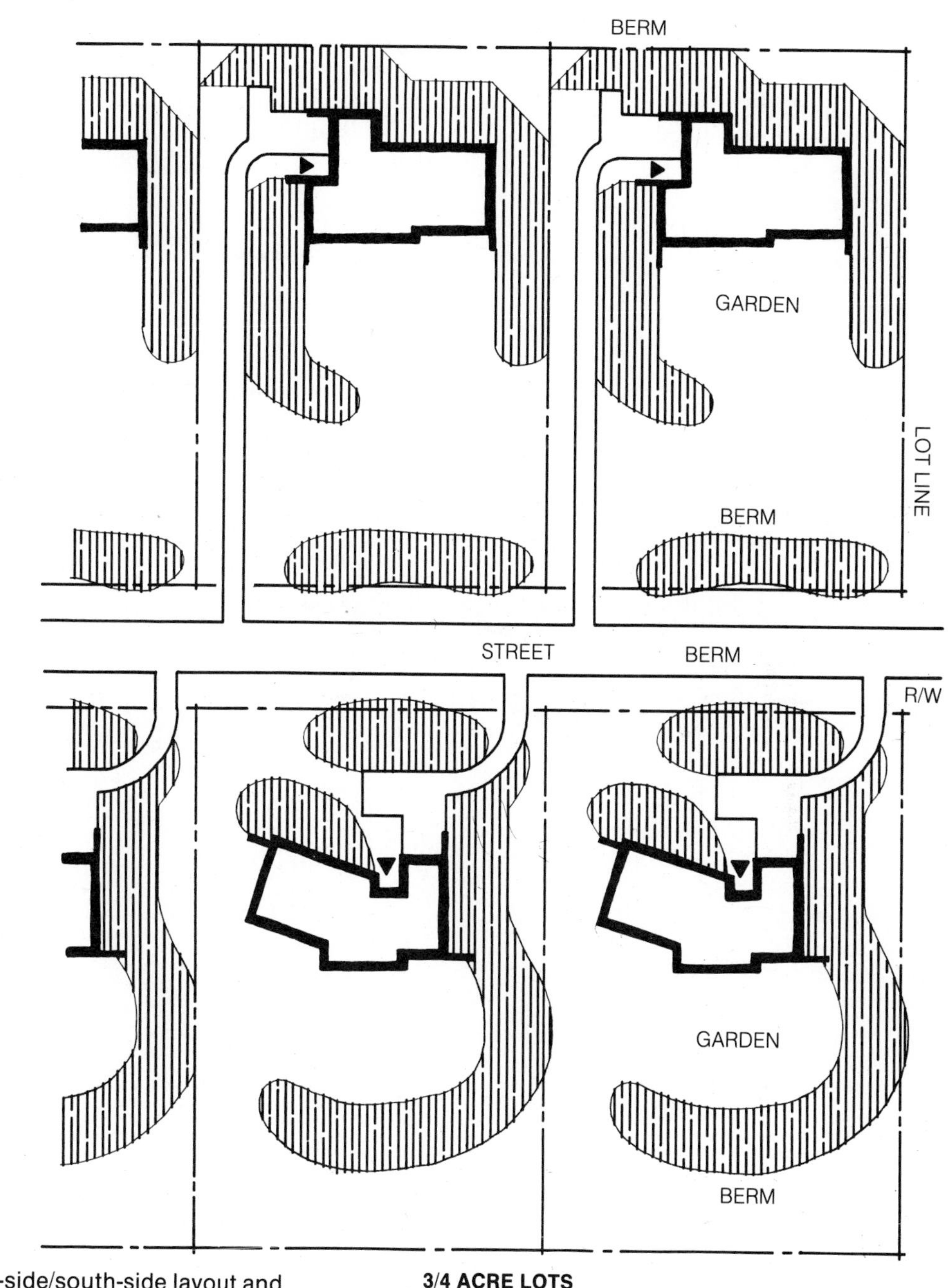

17–4. North-side/south-side layout and subdivision plan detail (on large lots).

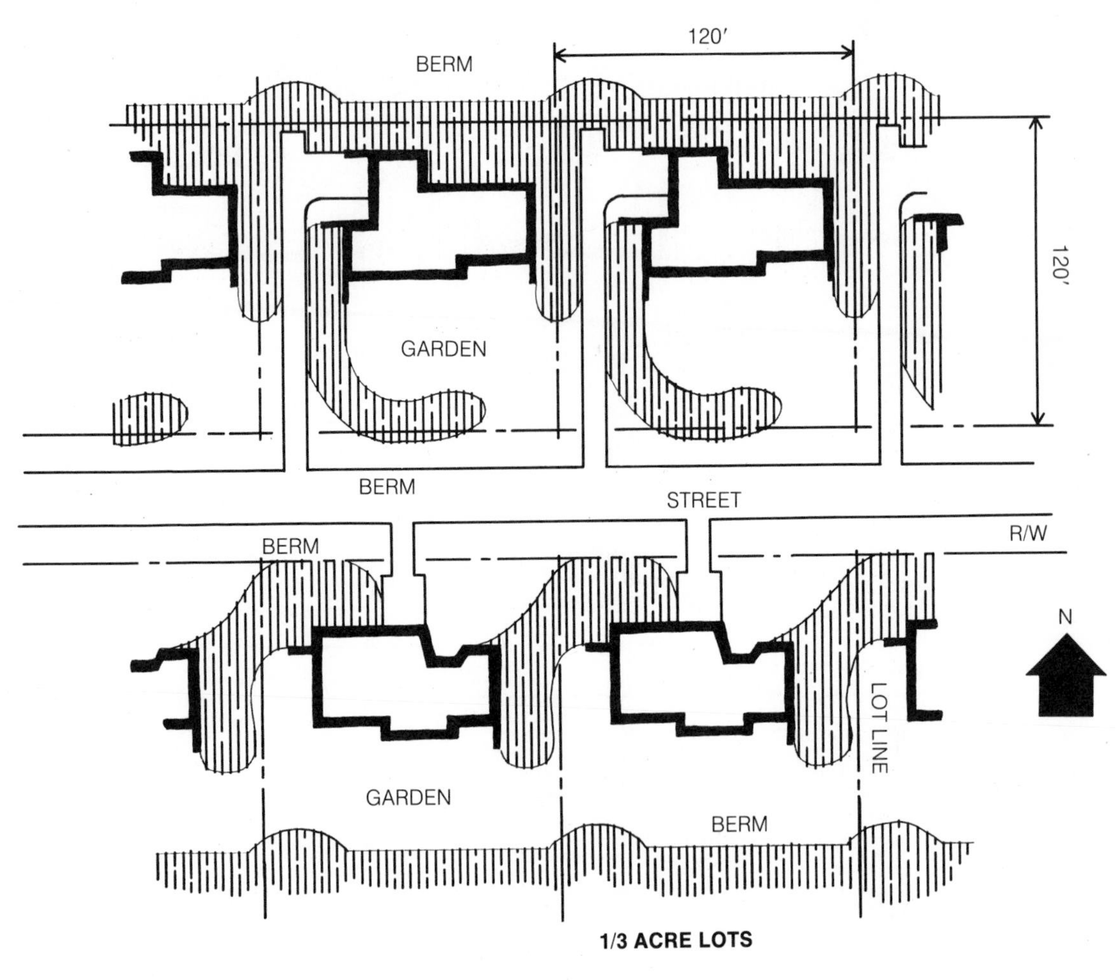

17–5. North-side/south-side layout and subdivision plan detail (on small lots).

With this arrangement, a long berm provides privacy for the outdoor living area. The berm separates the driveway from the garden and is extended across the front of the house. Privacy can be obtained with a berm 3 to 4 feet high, topped by a low planting screen; larger bushes, or a solid hedge, can be planted to establish a greater sense of enclosure. It is not necessary to carry the berm across the entire width of the lot, since only the outdoor living area opening off the dining room or kitchen must be screened from the road.

This arrangement will work if lot sizes are ample (see figure 17–4). The one-story house with a side entrance, however, will be too long to fit comfortably on a small lot, unless the house itself is relatively small. The 2,000-square-foot house will be almost 90 feet long, and a minimum lot width of 130 feet is necessary. This is much larger than the 75- to 90-foot frontage in the conventional subdivision. Building length and lot width can be reduced by introducing a two-story bedroom wing (figures 17–6, 17–7) of the type illustrated in figure 4–10.

Houses located on the north side of the street should be placed close to their rear property lines, since the ground behind the buildings is not useful, and only an access strip for maintenance is needed. This will leave much less room between the two rows of houses on each block than is provided in a conventional layout, where houses face each other across their backyards (figure 17–8). But because the one-story earth sheltered house is built into the ground and concealed with protective berms and planting screens, a greater, rather than a reduced, sense of openness and privacy is achieved with the more closely spaced ESSPS buildings.

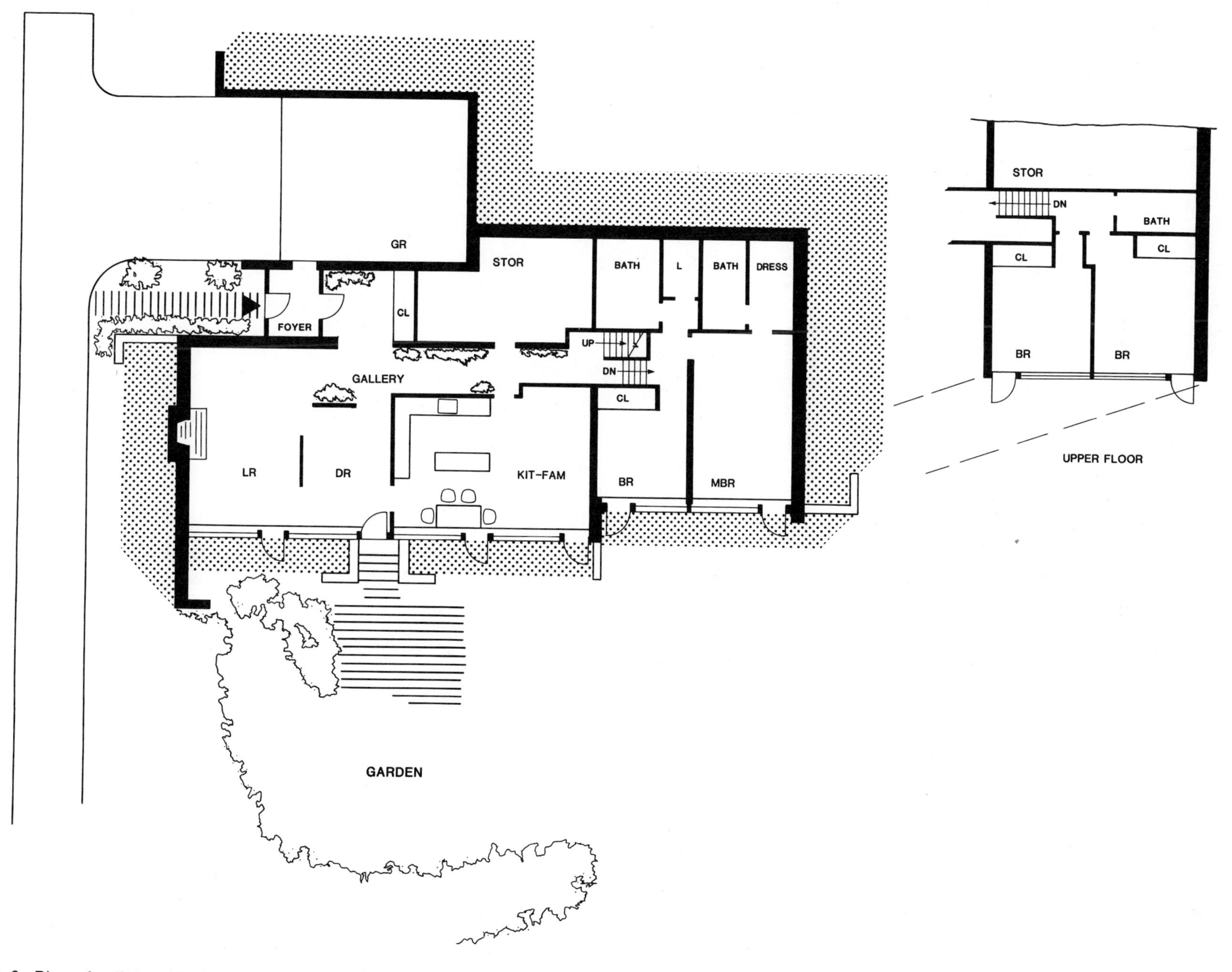

17–6. Plan of split-level house for a north-side lot.

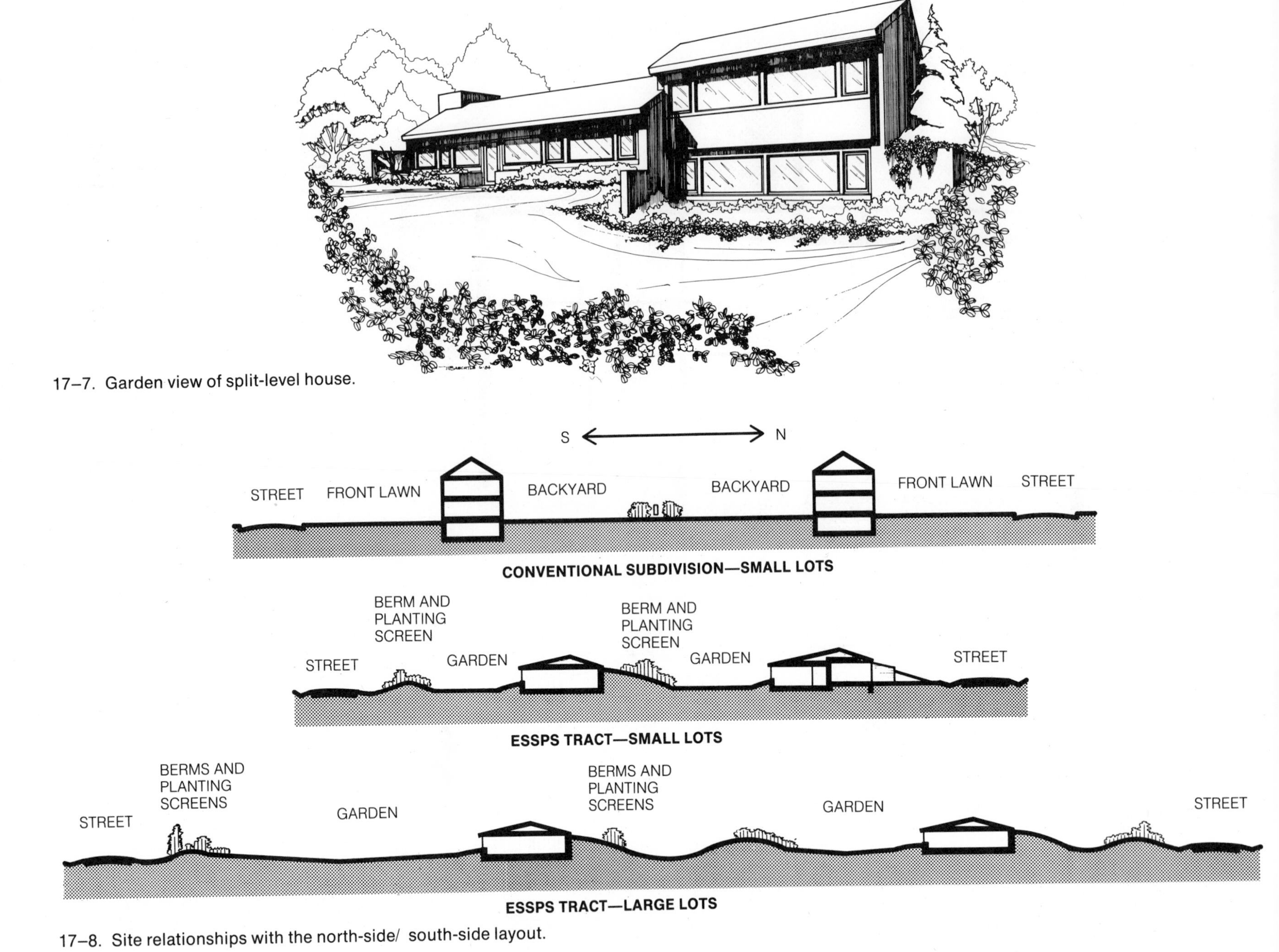

17–7. Garden view of split-level house.

17–8. Site relationships with the north-side/ south-side layout.

Orientation—Tracts with North/South Streets

Where streets run generally north and south, rather than east and west, an entirely different type of layout is required. Here, if the houses are to face south, they must be turned sideways along the street (figure 17–9). Every house will directly face its neighbor to the south, but will, again, face a closed building elevation that is hidden by a planted berm (figure 17–10). Since the backyard has now become a side yard, the south side of the house, with its large windows, and the outdoor living area must be protected. Here again berms and plantings can be used to shelter the lot. The protective berms can be carried across the whole width of the lot to screen the garden from both the street and the garage and driveway of the neighboring house.

Houses can also be placed at an angle to the street, creating a chevron layout, in which buildings on both sides of the street face up to 25 degrees away from due south (figure 17–11). The chevron plan opens the view past the berm sheltering the adjacent house, providing a combination of enclosed and open garden spaces. Winter gain from direct sunlight will be reduced by approximately 15 percent, but this will have little effect on the overall performance of the ESSPS house, which on clear days will continue to receive more solar energy than it can usefully store.

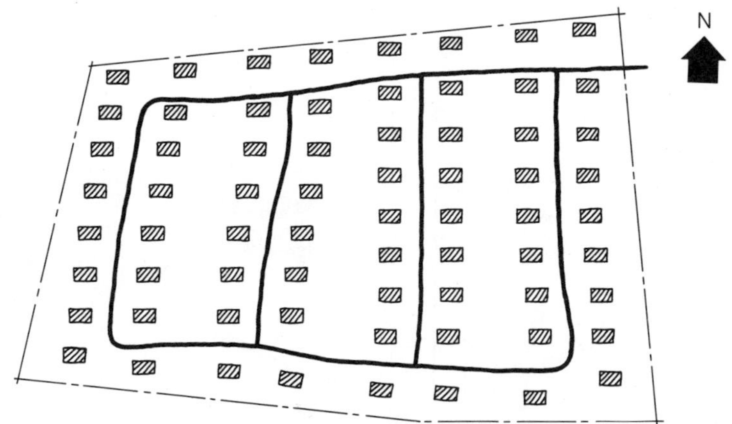

17–9. ESSPS subdivision plan with north/south streets.

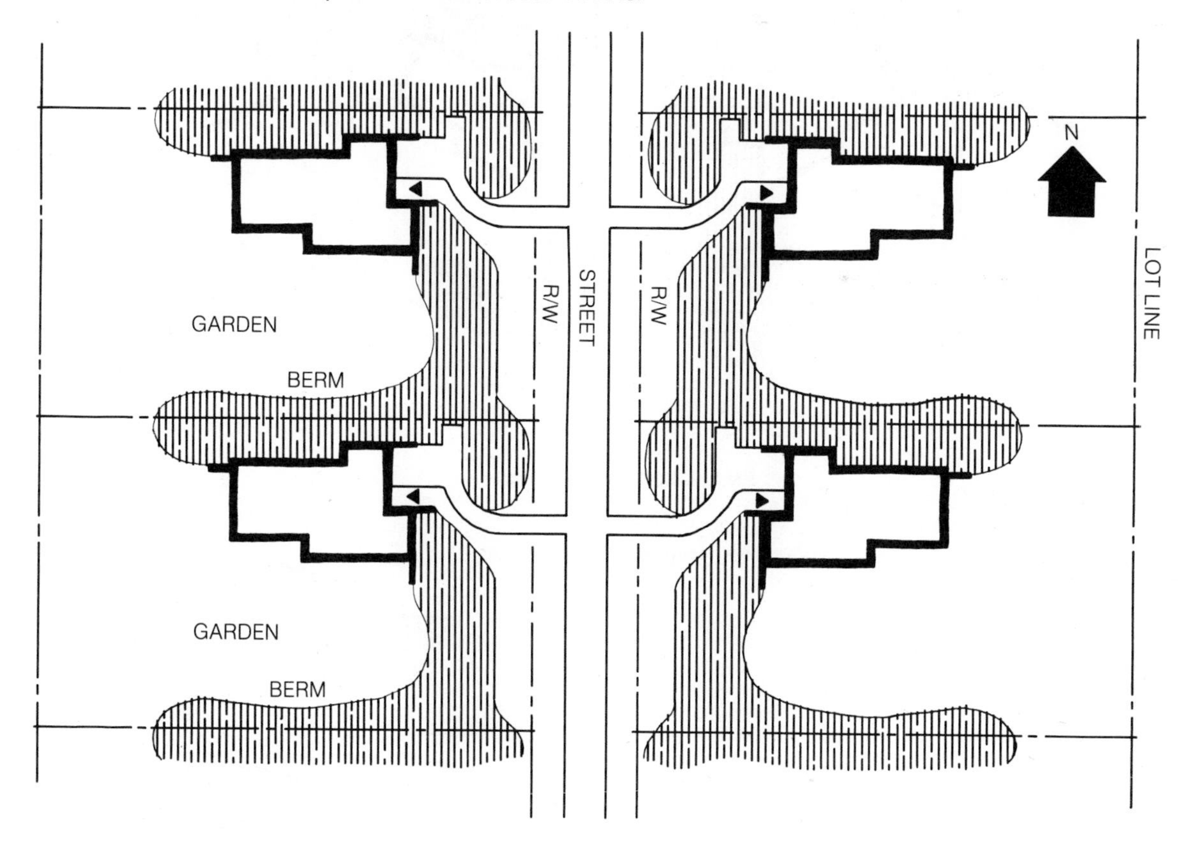

17–10. Detail of subdivision plans with sideways layout.

The chevron plan makes extremely efficient use of ground, providing large open yards, as well as short, economical driveways, and it allows a wide variety of designs for both the entrance area and the garden. The frontage needed for proper spacing of the houses will, again, be somewhat greater than in a conventional subdivision with small lots. The change from the front-yard/backyard plan, however, allows lot depth to be reduced, so that lot areas need not be increased. Where the chevron plan is used with lots of ½ acre or more, the traditional deep rectangular lot will require little alteration.

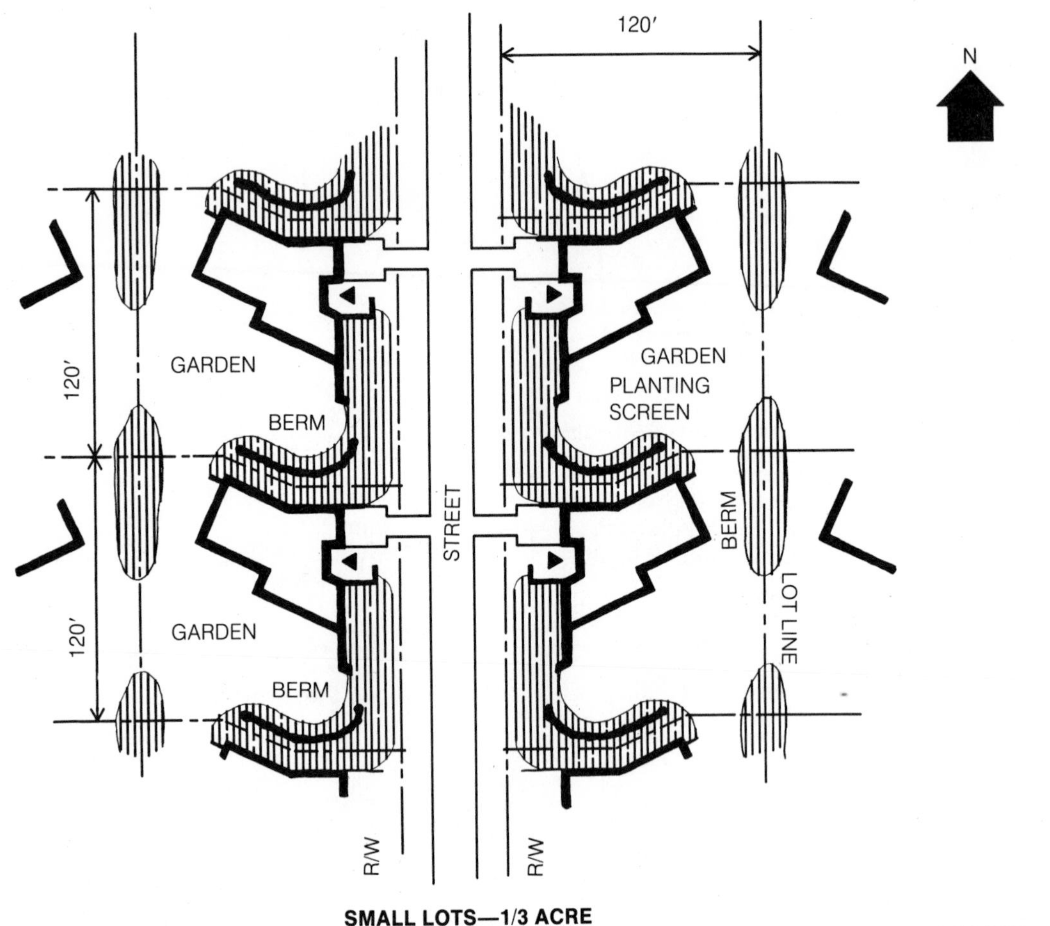

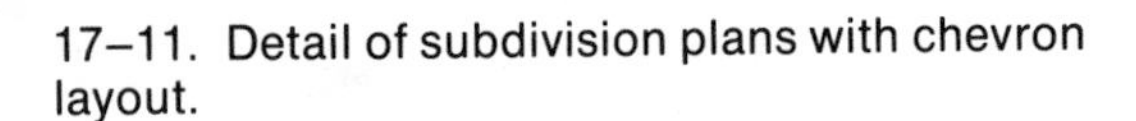
17–11. Detail of subdivision plans with chevron layout.

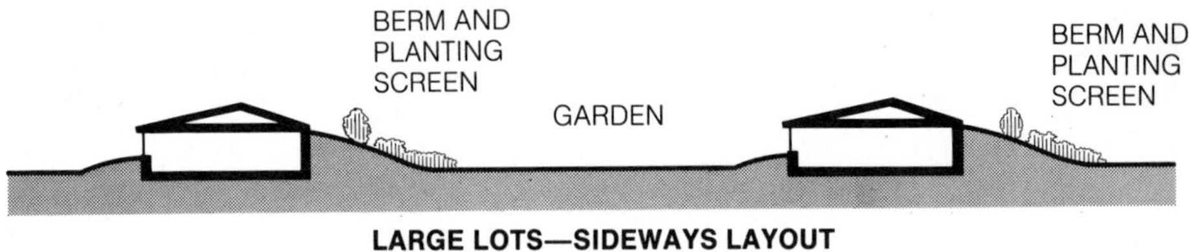

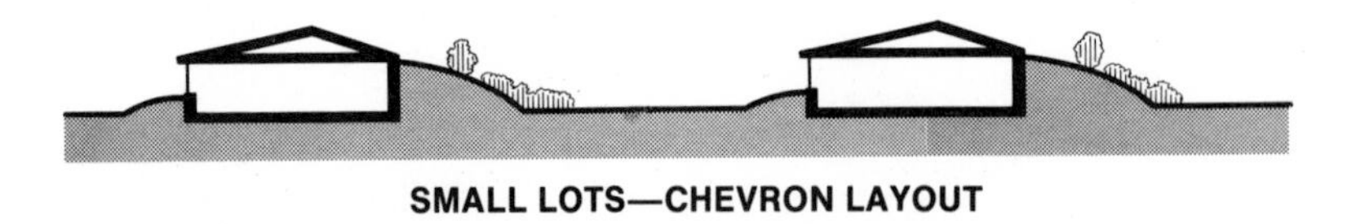

17–12. Site relationships with sideways and chevron layouts. Sections through garden areas.

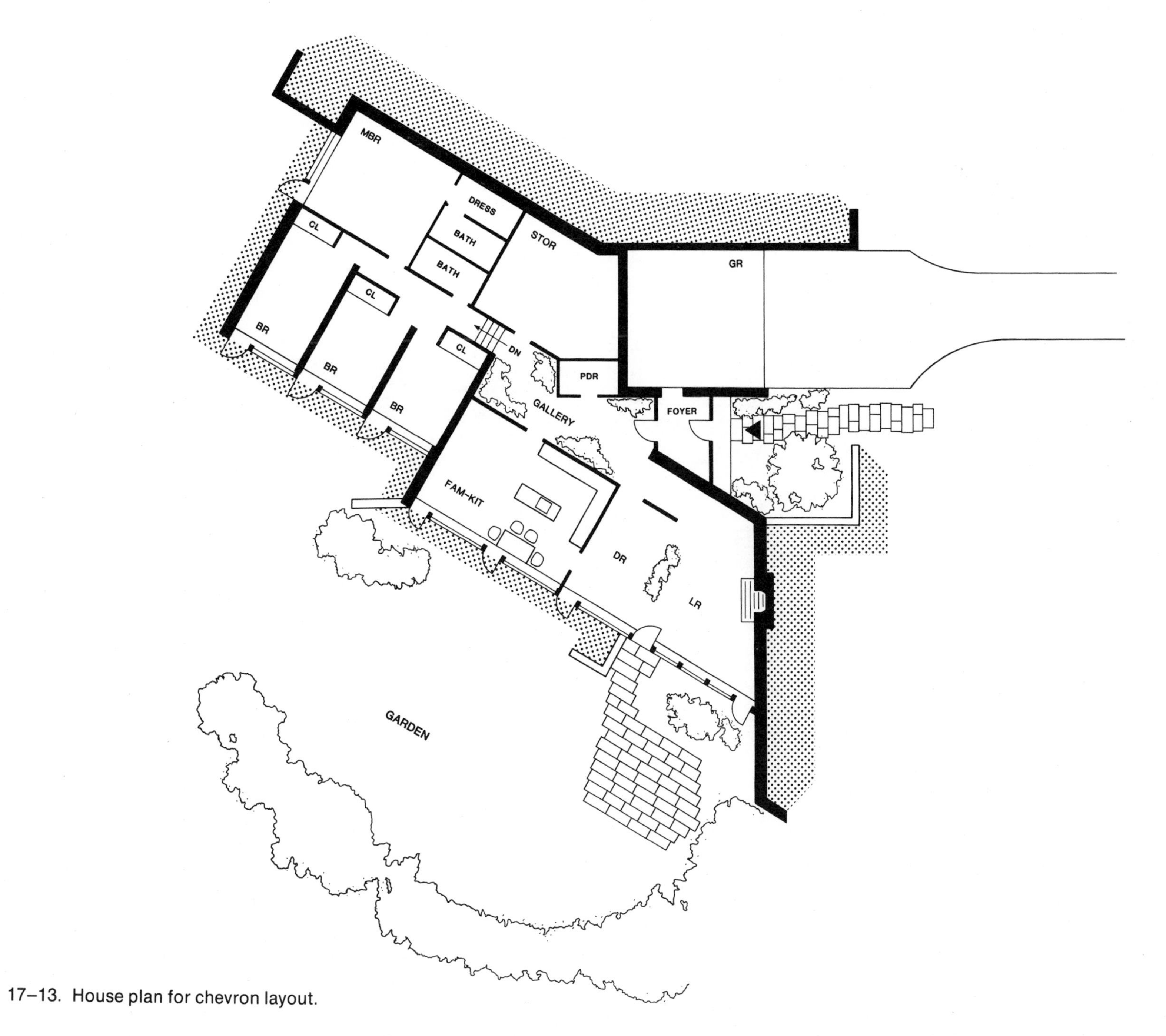

17–13. House plan for chevron layout.

17–14. View of house designed for chevron layout.

Flexible Street Layouts

Both the northside/southside lot arrangement on east/west streets, and the sideways arrangement on north/south streets allow a great deal of flexibility in street layout. Street direction can be turned as much as 25 degrees from the cardinal points of the compass without significantly affecting the passive solar performance of the houses in the subdivision. This allows the subdivision street plan to be drawn with varied, curvilinear layouts, and it will permit an efficient street and lot layout with irregular tract boundaries (figures 17–3, 17–9, 17–15). Where the street layout must be adapted to exceptionally irregular tract boundaries or unusually steep grades, still greater flexibility in the street plan may be needed. Figure 17–16 shows a design with a winding collector road, laid out to solve difficult grade conditions; a set of short cul-de-sac streets, oriented to ensure efficient passive solar performance, open off the collector road.

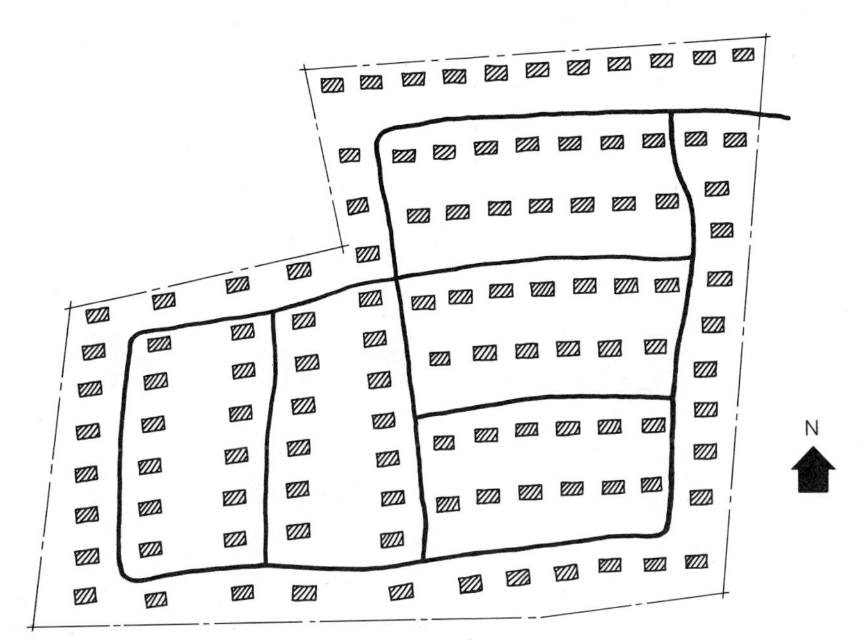

17–15. Subdivision plan combining north/south and east/west streets.

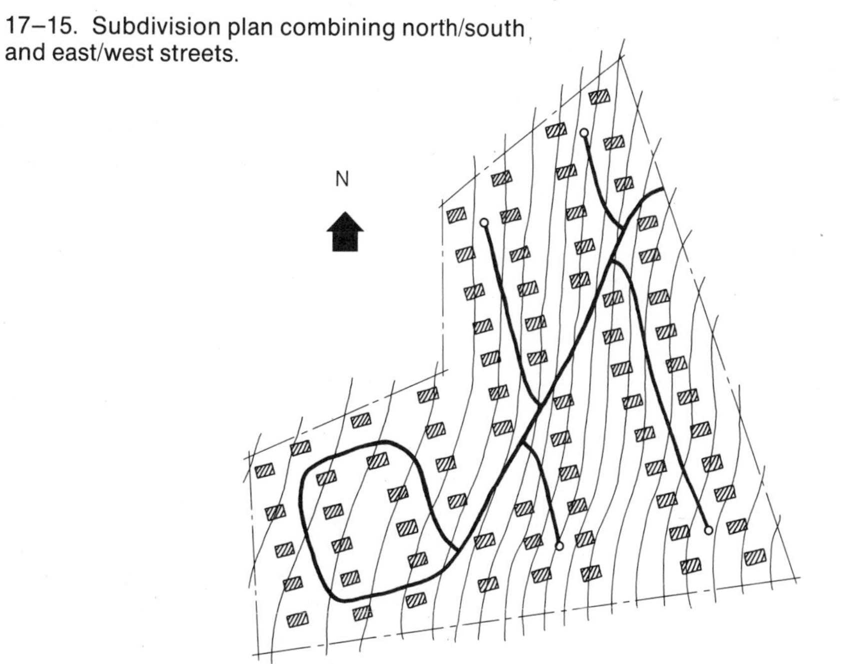

17–16. Spine and cul-de-sac subdivision plan.

18

Streetscape and Landscape

The Street Vista

Streets in the subdivision developed for the ESSPS house will bear little resemblance to the conventional suburban street. In the energy-efficient community, houses can be partially or entirely concealed from the street, and the streetscape will be dominated by natural and artificial earth forms, landscape elements, and open vistas. The tract planner no longer confronts the problem of the repetitive building facade, and the process of shaping the surface of the land itself becomes the source of a great variety of creative design opportunities.

Berms built along the street to provide privacy for courtyards and gardens can be kept low and set well back from the road so that the street appears open, or higher banks may be built close to the road, to suggest a country lane (figure 18–1). The tract designer can also vary the extent to which houses are visible from the street. Courtyard and entrance areas can be almost completely concealed, with only a driveway, cut through the bordering banks and berms, indicating the presence of a building. At inter-

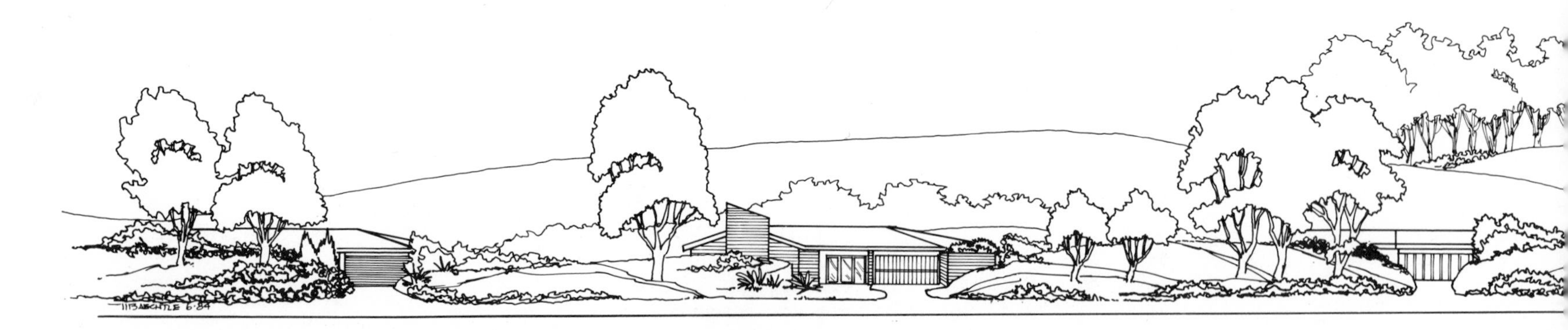

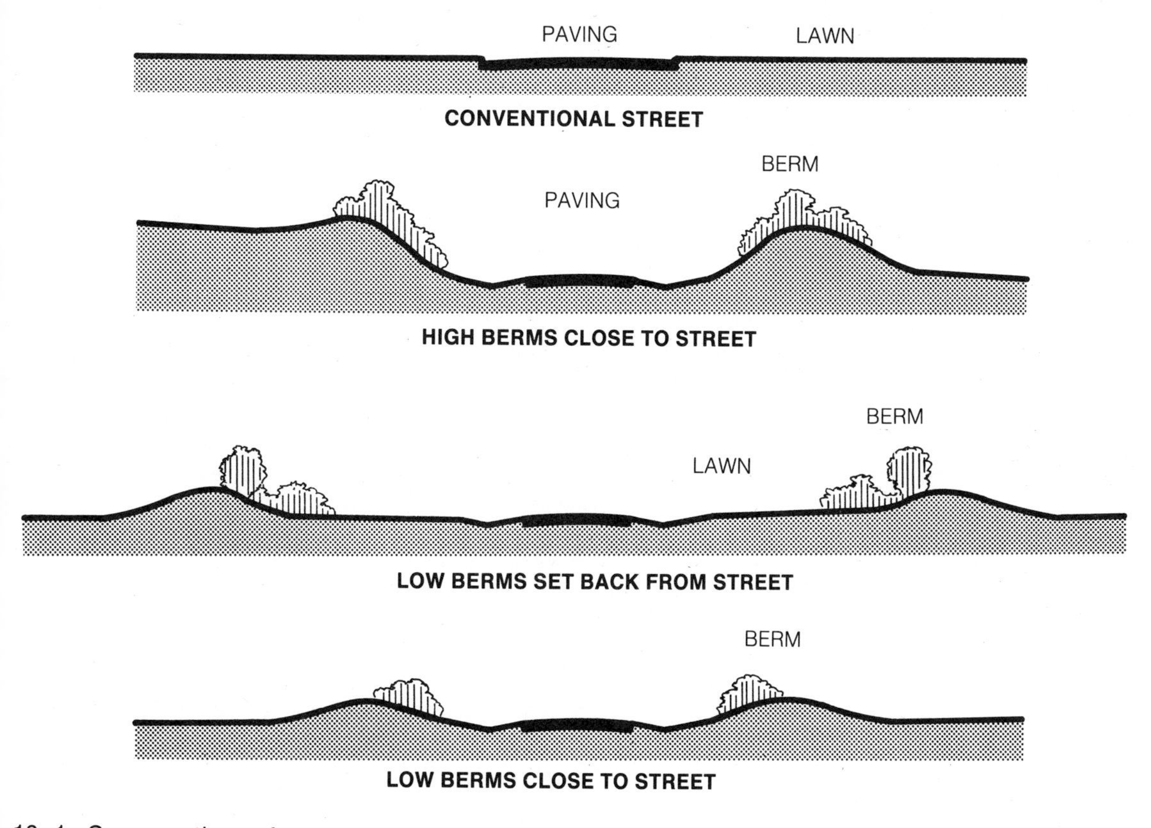

18–1. Cross sections of streets with different embankments.

18–2. Street elevations in the energy-efficient subdivision.

vals houses can be left visible from the street. Landscaping should complement the use of earth forms. Heavy, continuous planting can be established at some points along the road; sparser plantings will be appropriate wherever a wide vista opens over the bordering berms and the low roofs of the earth sheltered houses.

The appearance of the street will change naturally as the direction of the streets changes, and it will also vary according to the ground on which the subdivision is built. On an east/west street, all of the houses built on the north side of the street are set to the rear of their lots, and they will be largely concealed by their protective berms. Houses across the street have much shallower setbacks. On flat ground or gentle slopes their courtyards can be left open to the street, or concealed by earth forms and plantings in whatever pattern the designer chooses. If, however, land is moderately or steeply sloping, the houses on the south side of the street will be almost completely hidden from view behind the courtyard embankments cut into the hillside, so that here houses on both sides of the east/west street are concealed.

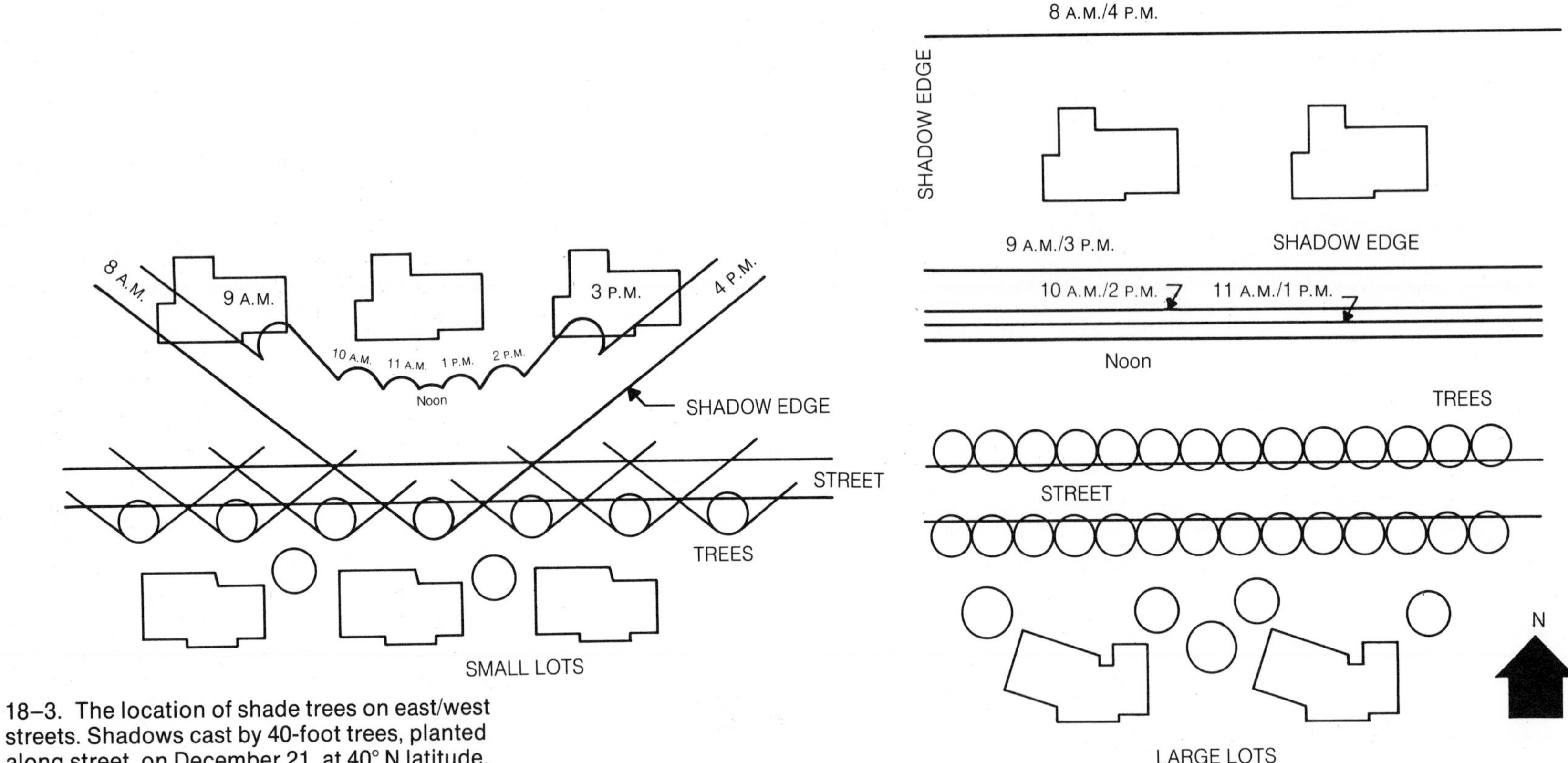

18–3. The location of shade trees on east/west streets. Shadows cast by 40-foot trees, planted along street, on December 21, at 40° N latitude.

A different pattern appears along the north/south streets. Houses built on these streets are turned sideways to the road, and their street elevations are formed of the end wall of the living space, the entrance, and the garage. Buildings on both sides of the street can be almost completely concealed, or, in places, open courtyards can be used, permitting the end elevation of the house to be wholly or partially visible from the street.

It may sometimes be desirable to vary the streetscape with still stronger emphasis on man-made forms. This can be done by building conventional one- and two-story superinsulated houses on a number of lots in the subdivision. If these are built at intervals, with the earth sheltered houses that separate them largely concealed by berms and planting screens, the subdivision street will resemble a country road on the edges of a small town, where generously spaced homes are built on large properties. Concealing most of the houses in the subdivision with earth forms and plantings will also give each of the visible buildings architectural significance; they will stand out in their individual settings, rather than form a part of a long row of closely spaced houses. The nonresidential buildings in the earth sheltered community—schools, churches and other public buildings, as well as the shopping centers—will also acquire much greater architectural importance, since they will be visible as isolated structures or groups of structures, rather than as large buildings surrounded by endless rows of small houses.

Landscaping and Placing Trees

Landscaping, as well as earth forms, introduces variety into the design of the streets in the earth sheltered community. Shade trees are part of this design, but they must be carefully placed. Shade trees are more difficult to place when all of the houses in a subdivision, rather than just a single house, must be planned for open winter sky and rel-

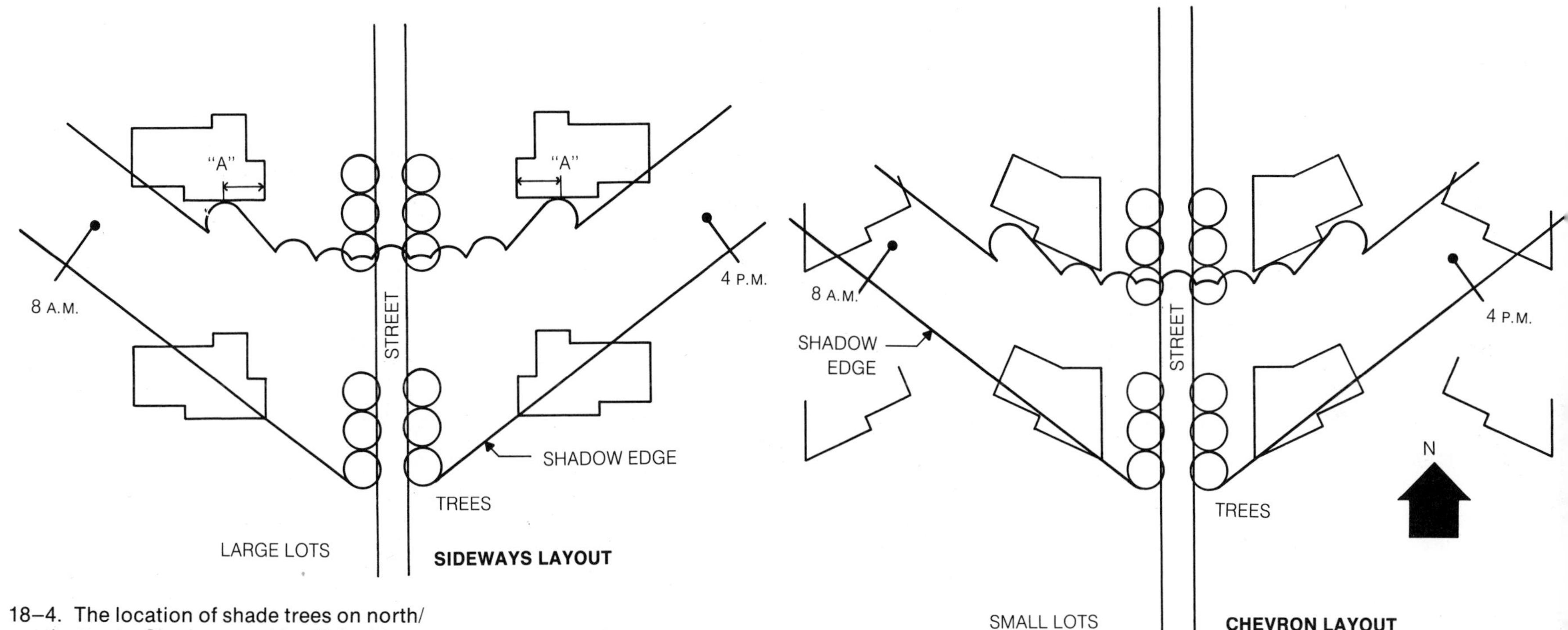

18–4. The location of shade trees on north/south streets. Shadows cast by 40-foot trees, planted along the street, on December 21, at 40° N latitude.

atively unobstructed passive gain. In the energy-efficient tract, trees planted on one lot will cut off excessive amounts of needed sunlight from the neighboring houses unless they are properly located.

Where the north side/south side layout is used, the south side of its east/west streets can be planted with a continuous row of medium-size shade trees, provided the lots are relatively large. Plantings on the north side of the street must be kept lower and trees and bushes on this side of the street should not exceed 20 to 25 feet at maturity. Still greater care must be taken when planting the east/west streets of a small lot development. Here, it will not be possible to avoid tree shadows on the windows of the houses on the north side of the street if shade trees reach a height of 35 to 40 feet. These shadows, however, will be cast by the early morning and late-afternoon sun, and if trees are spaced approximately 60 feet apart, loss of clear-day passive gain will be small, resulting in minimal impact on overall passive performance.

Shade trees on the north/south streets are planted in clusters at the end of each building. Tight grouping of major trees will prevent shadows from affecting passive gain, even where lots are small. If the tree clusters are extended beyond the ends of the houses, there will be small, but usually acceptable, shadow areas on the glass on the south side of the buildings.

In addition to the shade trees, smaller trees can be planted along the road where tall trees would cast shadows on the window walls of the houses. Plant masses on the banks will complete the planting scheme; they can include species such as the familiar rhododendron, which forms impressive planting screens, and clumps or rows of deciduous bushes.

The landscape design, with its combination of large and small trees and plant masses, can be used to create many different types of visual effects. Formal, parklike street settings can be designed with regular rows of trees standing on lawn areas bordering the road and extending back to low, nearly continuous berms. Woodland settings can also be established by reducing open lawn to a minimum along the street and establishing a mixture of trees, including deciduous shade trees and evergreens. The trees must be placed close together, and seedlings, rather than large saplings, will be planted. Contrasting with these two types of street design, an irregular landscape can also be created with scattered tree clumps and separate plant masses placed on berms and banks of varying height.

Still another type of landscape can be formed by placing heavy and continuous plant masses along the streets, so that berms and banks are covered with dense foliage. Here the subdivision streets will appear to wind through a heavily planted garden, within which are visible only glimpses of the buildings. Where the terrain permits open views, sparser planting will be desirable, both along the streets and on the interior of the lots. This will permit views down a valley into which the street is descending, or open the view from the valley floor to surrounding hills. The choices of visual effects will be determined by the characteristics of the site and its surrounding landscape. But because nature offers infinite variety, the change in emphasis from building elevations to landscape patterns will lead to more effective and original designs than are possible in the conventional suburban tract.

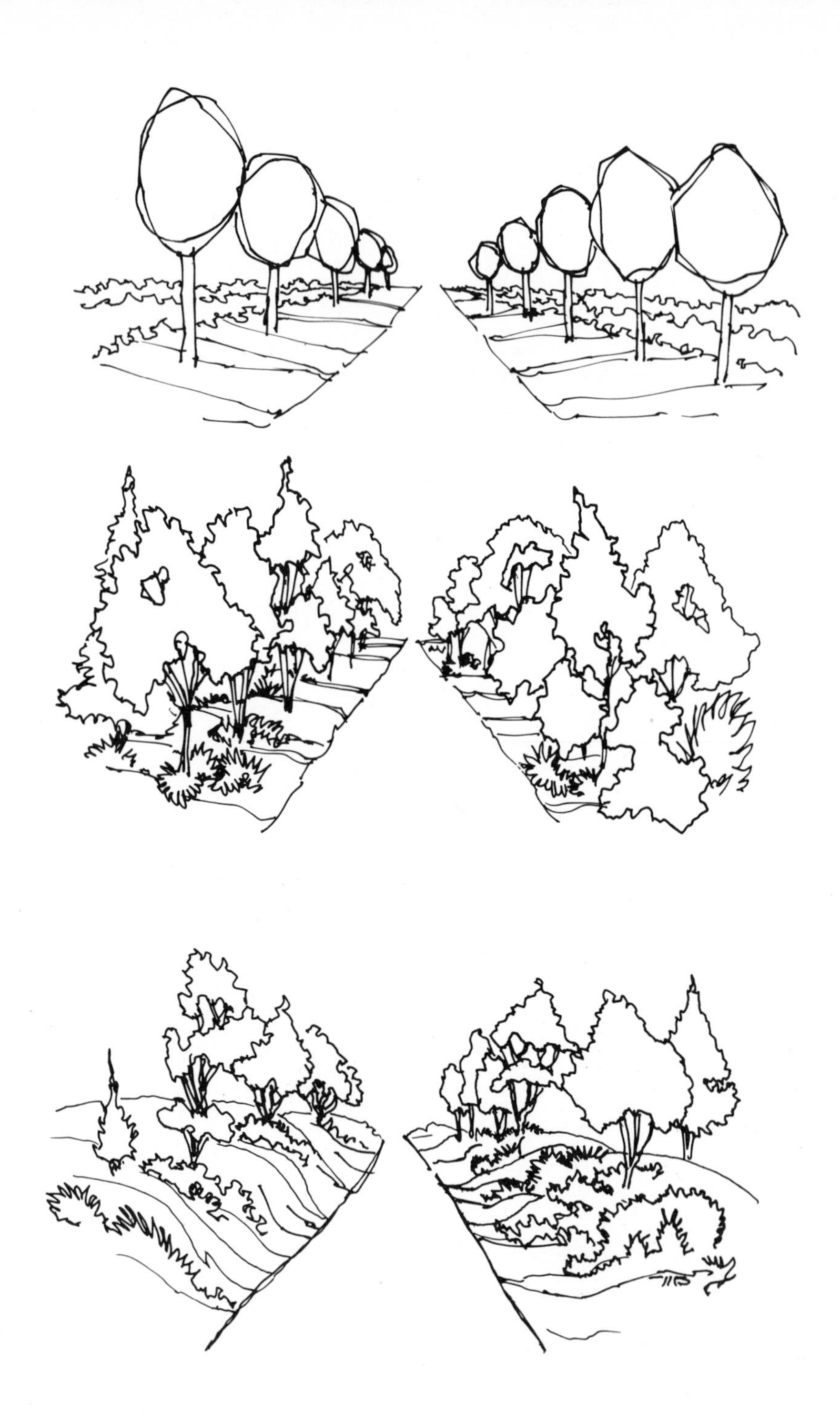

18–5. Streets and plantings in the energy-efficient subdivision.

APPENDICES

Appendix A

Below-Grade Heat Loss

Heat Loss Below Grade in Basement Walls

Heat loss, W/m²/°K				Depth	
Uninsulated	**R = 0.73**	**R = 1.47**	**R = 2.20**	**m**	**(ft)**
2.33	0.86	0.53	8	0–0.30	(0–1)
1.26	0.66	0.45	0.36	0.30–0.61	(1–2)
0.88	0.53	0.38	0.30	0.61–0.91	(2–3)
0.67	0.45	0.34	0.27	0.91–1.22	(3–4)
0.54	0.39	0.30	0.25	1.22–1.52	(4–5)
0.45	0.34	0.27	0.23	1.52–1.83	(5–6)
0.39	0.30	0.25	0.21	1.83–2.13	(6–7)

Path length through soil		Heat loss, Btu/h/ft²/°F			
m	**(ft)**	**Uninsulated**	**R = 4.17**	**R = 8.34**	**R = 12.51**
0.20	(0.68)	0.410	0.152	0.093	0.067
0.69	(2.27)	0.222	0.116	0.079	0.059
1.18	(3.88)	0.155	0.094	0.068	0.053
1.68	(5.52)	0.119	0.079	0.060	0.048
2.15	(7.05)	0.096	0.069	0.053	0.044
2.64	(8.65)	0.079	0.060	0.048	0.040
3.13	(10.28)	0.069	0.054	0.044	0.037

Heat Loss through Basement Floors (W/m²/K [Btu/h/ft²/°F])

Width of house, m				Depth of foundation wall below grade	
6	7.3	8.5	9.7	m	(ft)
0.18	0.16	0.15	0.13	1.50	(−5)
0.17	0.15	0.14	0.12	1.83	(−6)
0.16	0.15	0.13	0.12	2.10	(−7)

Width of house, ft			
20	24	28	32
0.032	0.029	0.026	0.023
0.030	0.027	0.025	0.022
0.029	0.026	0.023	0.021

Data for floor loss is for uninsulated slabs. The insulated ESSPS slab is 2½ feet to 7 feet below grade. It is treated as an uninsulated slab with an average depth of 7 feet.

SOURCE: ASHRAE Guide, 1981 ed.

Appendix B

Passive Solar Storage Capacity

The diurnal method of calculating passive solar storage distinguishes between surfaces in direct sunlight and surfaces receiving indirect sunlight. In the case of a surface in direct sunlight, diurnal heat capacity (dhc) is

$$dhc = sF_1 \text{ (Btu/°F/ft}^2\text{ surface)}. \qquad (1.1)$$

In this equation

$$s = \sqrt{(KqcP/2\pi)}. \qquad (1.2)$$

where K = thermal conductivity (Btu/ft/hr/°F)
q = density (lbs/ft³)
c = specific heat (Btu/lb/°F)
P = periodicity (taken as 24 hrs)

F_1 is a function of x, a dimensionless constant that varies with the thickness of the storage material.

$$x = L\sqrt{(\pi qc/PK)}. \qquad (1.3)$$

where L = thickness of storage material (ft)

Tabulated values of F_1 allow its value to be determined once x has been obtained. For a given component, storage capacity (ΔQ) equals the diurnal heat capacity × change in the surface temperature of the component as heat is stored × surface area.

$$\Delta Q = (dhc)\ (\Delta T)\ (A). \qquad (1.4)$$

where dhc = SF_1
A = area of storage surface (ft²)
T = diurnal swing in storage surface temperature (°F)

Room air temperature swing is used as an approximation for the change in surface temperature. This will underestimate the actual temperature swing of the storage element surface and lead to conservative design.

Surfaces that are not in direct sunlight are heated by convection of the room air and radiation from other interior surfaces, rather than by direct solar radiation. To compensate for this, for surfaces that are not in direct sunlight, the factor F_2 is added to eq. 1.1, and the diurnal heat capacity becomes:

$$dhc = sF_1F_2. \qquad (1.5)$$

F_2 is a function of x, which has been obtained previously, and of r, where

$$r = 2\pi sF_1/hP. \qquad (1.6)$$

The only new factor in this equation is h, the rate of convective heat transfer from the air to the surface of the storage material. This is commonly taken as 1.5 Btu/hr/sq/°F for still air conditions, with materials that have good heat transfer properties. If the surface is covered with a material having poor conductivity, h will have a lower value. Introducing F_2 will lower storage capacity of the storage elements.

Equations 1.3 and 1.6 give capacity for storage components with one exposed surface. They can be applied directly to storage in free-standing walls by dividing the wall into two equal slices.

Table B-1 Dimensionless Thickness X

r	0.0	.2	.4	.6	.8	1.0	1.2	1.4	1.6	1.8	2.0	3.0
.00	1.000	1.000	1.000	1.000	1.000	.999	.999	.999	.999	.999	.999	.999
.05	.999	.997	.994	.987	.980	.974	.969	.967	.965	.965	.964	.965
.10	.995	.992	.985	.973	.960	.948	.939	.934	.932	.931	.930	.932
.15	.989	.985	.974	.957	.938	.922	.910	.903	.900	.898	.898	.900
.20	.981	.976	.961	.940	.916	.896	.882	.873	.869	.867	.867	.869
.25	.970	.964	.947	.922	.894	.871	.854	.845	.840	.838	.837	.840
.30	.958	.951	.931	.902	.872	.846	.828	.817	.812	.810	.810	.812
.35	.944	.936	.914	.883	.849	.821	.802	.791	.786	.783	.783	.786
.40	.928	.920	.896	.863	.827	.798	.778	.766	.761	.758	.758	.761
.45	.912	.903	.878	.842	.805	.775	.755	.743	.737	.734	.734	.737
.50	.894	.885	.859	.822	.784	.753	.732	.720	.714	.712	.711	.715
.60	.857	.848	.820	.782	.743	.711	.690	.678	.672	.670	.669	.673
.70	.819	.809	.781	.743	.704	.673	.652	.640	.634	.632	.632	.635
.80	.781	.771	.743	.706	.668	.637	.617	.606	.600	.598	.597	.601
.90	.743	.734	.707	.671	.634	.605	.585	.575	.569	.567	.566	.569
1.00	.707	.698	.672	.638	.602	.575	.556	.546	.541	.539	.538	.541
1.10	.673	.664	.640	.607	.573	.547	.530	.520	.515	.513	.512	.515
1.20	.640	.632	.609	.578	.547	.522	.505	.496	.491	.489	.489	.492
1.30	.610	.602	.581	.551	.522	.498	.483	.474	.469	.468	.467	.470
1.40	.581	.574	.554	.527	.499	.477	.462	.454	.449	.448	.447	.450
1.50	.555	.548	.529	.504	.478	.457	.443	.435	.431	.429	.429	.431
1.60	.530	.524	.506	.482	.458	.438	.425	.418	.414	.413	.412	.414
1.70	.507	.501	.485	.463	.440	.421	.409	.402	.398	.397	.397	.398
1.80	.486	.480	.465	.444	.423	.405	.394	.387	.384	.382	.382	.384
1.90	.466	.461	.447	.427	.407	.390	.379	.373	.370	.369	.368	.370
2.00	.447	.443	.429	.411	.392	.376	.366	.360	.357	.356	.356	.357
2.20	.414	.410	.398	.382	.365	.351	.342	.337	.334	.333	.333	.334
2.40	.385	.381	.371	.357	.342	.329	.321	.316	.314	.313	.312	.314
2.60	.359	.356	.347	.334	.321	.310	.302	.298	.295	.295	.294	.296
2.80	.336	.334	.326	.314	.302	.292	.285	.281	.279	.278	.278	.279
3.00	.316	.314	.307	.296	.286	.276	.270	.267	.265	.264	.264	.265
3.20	.298	.296	.290	.280	.271	.262	.257	.253	.252	.251	.251	.252
3.40	.282	.280	.274	.266	.257	.250	.244	.241	.240	.239	.239	.240
3.60	.268	.266	.261	.253	.245	.238	.233	.230	.229	.228	.228	.229
3.80	.254	.253	.248	.241	.234	.227	.223	.220	.219	.219	.218	.219
4.00	.243	.241	.237	.230	.223	.218	.214	.211	.210	.209	.209	.210
4.50	.217	.216	.212	.207	.201	.197	.193	.191	.190	.190	.190	.190
5.00	.196	.195	.192	.188	.183	.179	.176	.175	.174	.173	.173	.174

SOURCE: "Passive Solar Design Analysis." In *Passive Solar Design Handbook*. New York: Van Nostrand Reinhold Company, 1984.

Passive storage capacity for the 2,000-square-foot ESSPS house is calculated using the diurnal method. Direct sunlight striking the storage surfaces is not taken into account, and only the equation $dhc = F_1F_2$ is used. Values of s, x, and r are derived from equations 1.2, 1.3, and 1.6. Values for F_1 and F_2 are tabulated in tables B-1 and B-2.

Gypsum Board

All walls and ceilings in the house are covered with ½-inch gypsum board (drywall). Gypsum board area is 6,200 square feet. For gypsum board:

K = 0.25 Btu/ft/°F/hr	s = 4.23
q = 78 lbs/cu ft	x = 0.130
c = 0.24 Btu/lb/°F	F_1 = 0.184
L = 1/24 ft	r = 0.136
	F_2 = 0.589

then $sF_1F_2 = 0.458$

$\Delta Q = dhc\ (A)\ (\Delta T)$ where A = 6,200 sq ft
$\Delta Q = 22{,}716$ Btu $\Delta T = 8°$ F

Slab Areas Covered with Ceramic Tile, with Vinyl, and Left Exposed

These areas total 1,150 square feet. h = 1.5. Slab thickness is increased to 6 inches.

K = 1.0 Btu/ft/°F/hr	s = 10.71
q = 144 lbs/cu ft	x = 1.33
c = 0.21 Btu/lb/°F	F_1 = 1.13
L = 0.67 ft	r = 2.12
	F_2 = 0.361

then $sF_1F_2 = 4.31$

$\Delta Q = dhc\ (A)\ (\Delta T)$ where A = 1,150 sq ft
$\Delta Q = 39{,}652$ Btu T = 8° F

Slab—Carpeted Areas

The carpeted area of the slab is 1,330 square feet. h in eq. 1.6 becomes 0.5 to account for reduced rate of heat transfer through the carpet. Slab thickness is 4 inches.

K = 1.0 Btu/ft/°F/hr	s = 10.71
q = 144 lbs/cu ft	x = 0.66
c = 0.21 Btu/lb/°F	F_1 = 0.88
L = 0.33 ft	r = 4.96
	F_2 = 0.194

then $sF_1F_2 = 1.84$

$\Delta Q = dhc\,(A)\,(\Delta T)$ where A = 1,330
Q = 19,578 Btu T = 8° F

Block Storage Walls

Sixty-six lineal feet of walls are built as part of the system. The walls are 8-inch block, solidly filled. Walls are exposed on both sides and are treated as double 4-inch walls, with a total exposed surface area of 672 square feet.

K = 1.0 Btu/ft/°F/hr	s = 10.71
q = 144 lbs/cu ft	x = 0.66
c = 0.21 Btu/lb/°F	F_1 = 0.88
L = 0.33 ft	r = 2.12
	F_2 = 0.388

then $sF_1F_2 = 3.67$

$\Delta Q = dhc\,(A)\,(\Delta T)$ where A = 672 sq ft
Q = 23,500 Btu ΔT = 8° F

Phase Change Materials

Twenty-six 2,000-Btu-capacity tubes installed in plenum in the mechanical room. Phase change occurs at 67° F.

Table B-2 **Function F_1**

x	F_1	x	F_1	x	F_1	x	F_1	x	F_1	x	F_1
		.30	.423	.60	.816	.90	1.076	1.40	1.122	2.00	1.024
.01	.014	.31	.437	.61	.828	.91	1.081	1.42	1.118	2.05	1.019
.02	.028	.32	.451	.62	.839	.92	1.086	1.44	1.115	2.10	1.015
.03	.042	.33	.465	.63	.850	.93	1.091	1.46	1.111	2.15	1.011
.04	.057	.34	.479	.64	.861	.94	1.095	1.48	1.107	2.20	1.008
.05	.071	.35	.493	.65	.872	.95	1.100	1.50	1.104	2.25	1.005
.06	.085	.36	.506	.66	.883	.96	1.104	1.52	1.100	2.30	1.002
.07	.099	.37	.520	.67	.894	.97	1.107	1.54	1.096	2.35	1.000
.08	.113	.38	.534	.68	.904	.98	1.111	1.56	1.092	2.40	.999
.09	.127	.39	.548	.69	.914	.99	1.114	1.58	1.089	2.45	.997
.10	.141	.40	.561	.70	.924	1.00	1.117	1.60	1.085	2.50	.996
.11	.156	.41	.575	.71	.934	1.02	1.123	1.62	1.081	2.55	.995
.12	.170	.42	.588	.72	.943	1.04	1.128	1.64	1.077	2.60	.995
.13	.184	.43	.602	.73	.952	1.06	1.132	1.66	1.074	2.65	.994
.14	.198	.44	.615	.74	.961	1.08	1.136	1.68	1.070	2.70	.994
.15	.212	.45	.628	.75	.970	1.10	1.138	1.70	1.067	2.75	.994
.16	.226	.46	.642	.76	.979	1.12	1.140	1.72	1.063	2.80	.994
.17	.240	.47	.655	.77	.987	1.14	1.142	1.74	1.060	2.85	.994
.18	.254	.48	.668	.78	.996	1.16	1.143	1.76	1.057	2.90	.995
.19	.269	.49	.681	.79	1.004	1.18	1.143	1.78	1.053	2.95	.995
.20	.283	.50	.694	.80	1.011	1.20	1.143	1.80	1.050	3.00	.995
.21	.297	.51	.707	.81	1.019	1.22	1.142	1.82	1.047	3.25	.997
.22	.311	.52	.719	.82	1.026	1.24	1.141	1.84	1.044	3.50	.999
.23	.325	.53	.732	.83	1.033	1.26	1.140	1.86	1.041	3.75	1.000
.24	.339	.54	.744	.84	1.040	1.28	1.138	1.88	1.039	4.00	1.000
.25	.353	.55	.757	.85	1.047	1.30	1.136	1.90	1.036	4.25	1.000
.26	.367	.56	.769	.86	1.053	1.32	1.133	1.92	1.033	4.50	1.000
.27	.381	.57	.781	.87	1.059	1.34	1.131	1.94	1.031	4.75	1.000
.28	.395	.58	.793	.88	1.065	1.36	1.128	1.96	1.029	5.00	1.000
.29	.409	.59	.805	.89	1.071	1.38	1.125	1.98	1.026		

SOURCE: "Passive Solar Design Analysis." In *Passive Solar Design Handbook.* New York: Van Nostrand Reinhold Company, 1984.

Appendix C

Winter Solar Gain

Average Daily Global Radiation on a South-facing Surface

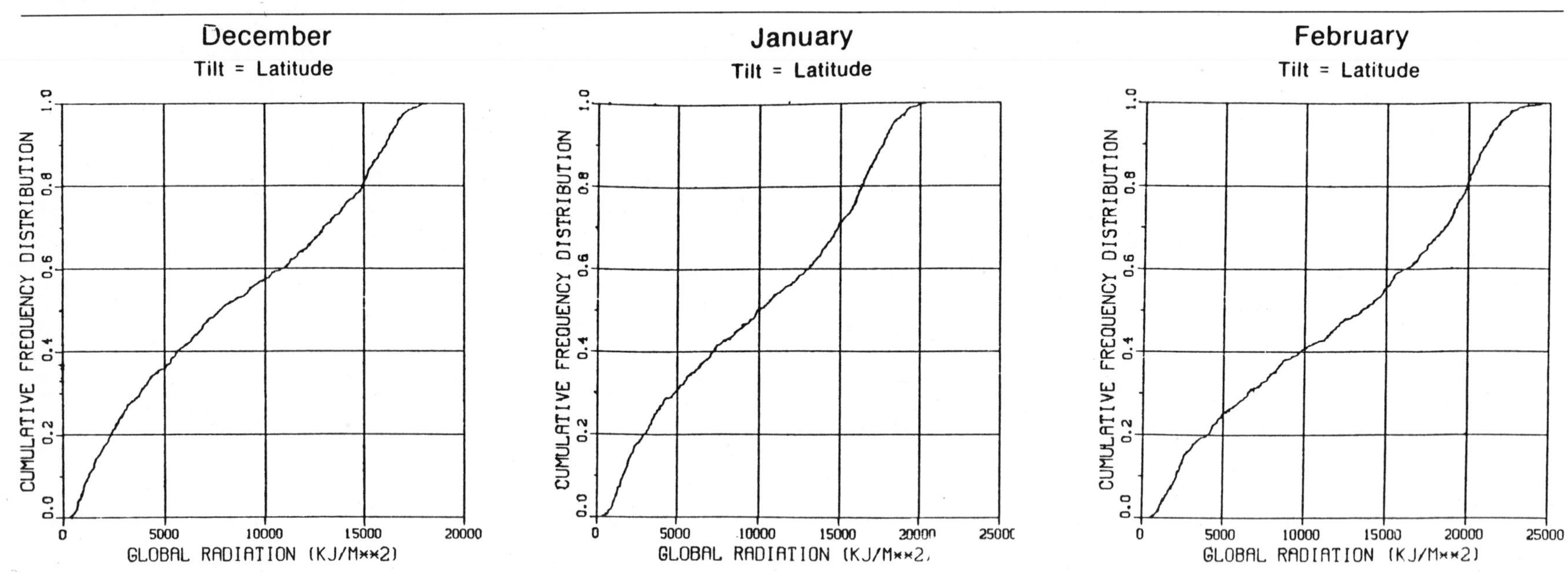

Radiation Distribution by Days
Washington, D.C./Sterling, VA

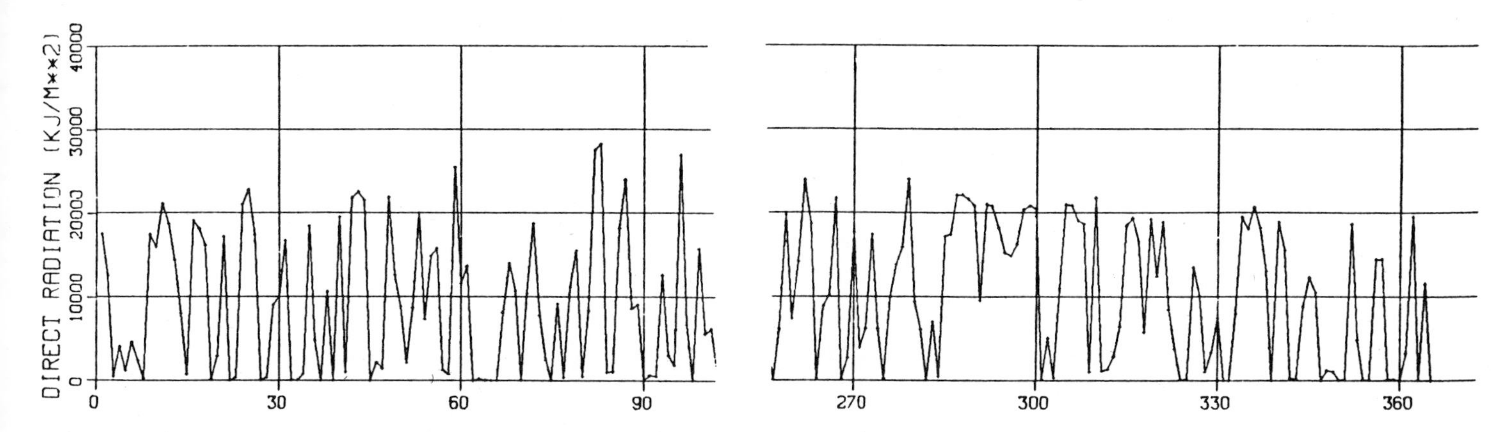

Daily Direct Normal Solar Radiation Year 1973 Washington, D.C./Sterling, VA

SOURCE: *Solar Radiation Atlas of the United States.* Solar Energy Institute, Washington, D.C., 1981.

Average Hourly Values of Direct Normal Solar Radiation

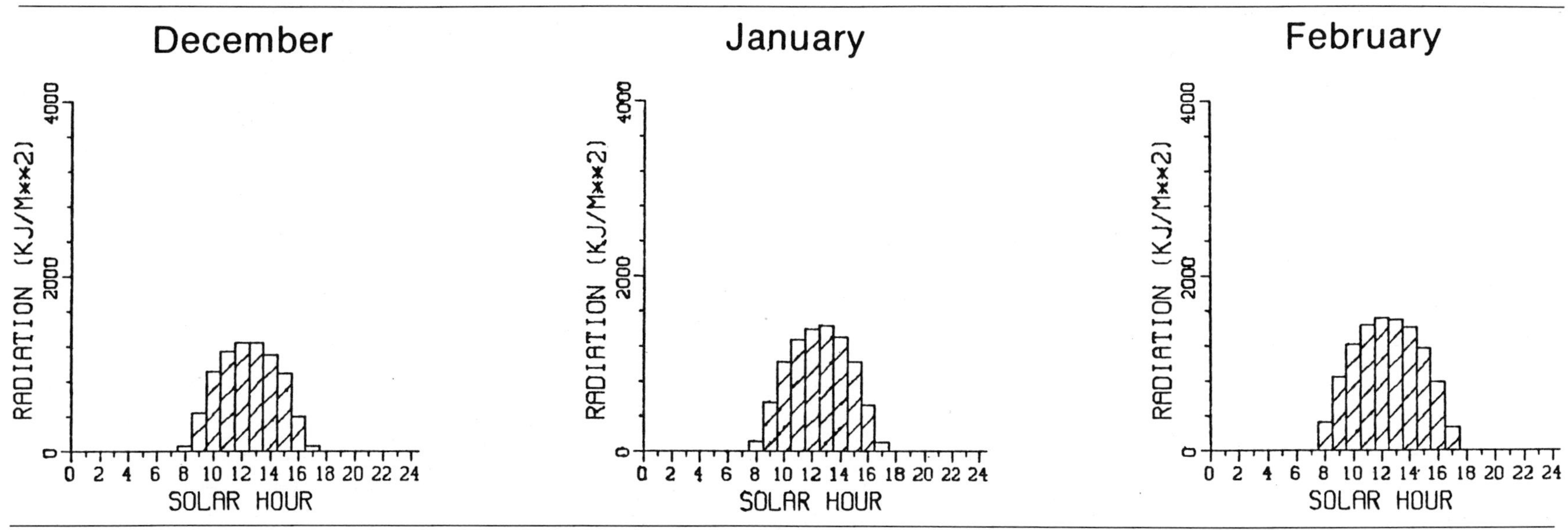

Washington, D.C./Sterling VA

SOURCE: *Solar Radiation Atlas of the United States.* Solar Energy Institute, Washington, D.C., 1981.

Appendix D

Mortgage Payments and Cash-flow Balance

Energy Savings and Carrying Costs for the Philadelphia Area ESSPS House

		Energy Costs*			Carrying Costs			
	Year	1985 Conventional House	ESSPS House	ESSPS Energy Cost Saving	Extra ESSPS Mortgage Payment**	Federal Income Tax Credit	Extra ESSPS Carrying Cost	ESSPS Annual Saving
6% 25-year mortgage, 0% energy price inflation	1	$ 1,404	$ 330	$ 1,074	$ 949	$183	$ 766	$ 308
	5	1,404	330	1,074	949	167	782	292
	10	1,404	330	1,074	949	143	806	268
	15	1,404	330	1,074	949	110	839	235
	20	1,404	330	1,074	949	66	883	191
	25	1,404	330	1,074	949	7	942	132
	26	1,404	330	1,074	0	0	0	1,074
12% 25-year mortgage, 0% energy price inflation	1	$ 1,404	$ 330	$ 1,074	$1,550	$367	$1,183	$ (109)
	5	1,404	330	1,074	1,550	353	1,197	(123)
	10	1,404	330	1,074	1,550	326	1,224	(150)
	15	1,404	330	1,074	1,550	278	1,272	(198)
	20	1,404	330	1,074	1,550	186	1,364	(290)
	25	1,404	330	1,074	1,550	21	1,529	(455)
	26	1,404	330	1,074	0	0	0	1,074

	Year	Energy Costs* 1985 Conventional House	ESSPS House	ESSPS Energy Cost Saving	Carrying Costs Extra ESSPS Mortgage Payment**	Federal Income Tax Credit	Extra ESSPS Carrying Cost	ESSPS Annual Saving
12% 25-year mortgage, 3% energy price inflation	1	$ 1,404	$ 330	$ 1,074	$1,550	$367	$1,183	$ (109)
	5	1,580	371	1,209	1,550	353	1,197	12
	10	1,832	431	1,401	1,550	326	1,224	177
	15	2,124	499	1,625	1,550	278	1,272	353
	20	2,462	579	1,883	1,550	186	1,364	519
	25	2,854	671	2,183	1,550	21	1,529	654
	26	2,940	691	2,249	0	0	0	2,249
12% 25-year mortgage, 5% energy price inflation	1	$ 1,404	$ 330	$ 1,074	$1,550	$367	$1,183	$ (109)
	5	1,706	401	1,305	1,550	353	1,197	108
	10	2,178	512	1,666	1,550	326	1,224	442
	15	2,780	653	2,127	1,550	278	1,272	855
	20	3,548	834	2,714	1,550	186	1,364	1,350
	25	4,528	1,064	3,464	1,550	21	1,529	1,935
	26	4,757	1,117	3,637	0	0	0	3,637
12% 25-year mortgage, 8% energy price inflation	1	$ 1,404	$ 330	$ 1,074	$1,550	$367	$1,183	$ (109)
	5	1,910	449	1,461	1,550	353	1,197	264
	10	2,806	660	2,146	1,550	326	1,224	922
	15	4,124	969	3,155	1,550	278	1,272	1,883
	20	6,059	1,424	4,635	1,550	186	1,364	3,271
	25	8,903	2,093	6,810	1,550	21	1,529	5,281
	26	9,615	2,260	7,355	0	0	0	7,355
12% 25-year mortgage, 10% energy price inflation	1	$ 1,404	$ 330	$ 1,074	$1,550	$367	$1,183	$ (109)
	5	2,056	483	1,573	1,550	353	1,197	376
	10	3,311	778	2,533	1,550	326	1,224	1,309
	15	5,332	1,253	4,079	1,550	278	1,272	2,807
	20	8,586	2,018	6,568	1,550	186	1,364	5,204
	25	13,829	3,250	10,579	1,550	21	1,529	9,050
	26	15,212	3,575	11,637	0	0	0	11,637

* Space heating, hot-water supply, air-conditioning
** After using renewable energy credit.

Appendix E

ESSPS Design for a Severe Winter Climate

Performance of the ESSPS house in a region with severe winter heating conditions is illustrated here by calculating performance for a home located in Minneapolis, Minnesota (45° N lat.). Minneapolis has a heating season with 8,380 degree days; the summer cooling period is 641 hours. Sunlight availability is about average for the northern United States. Average daily temperatures in the winter months are 12° F to 18° F.

For these conditions, the basic ESSPS house described in Part I is altered by going from R-34 to R-47 walls and by increasing passive solar storage capacity from 165,000 Btu to 213,000 Btu. The extra storage capacity is obtained by adding more tubes of phase change heat storage material to the plenum located in the mechanical room.

Table E-1 gives daily heat loss for the coldest month, January.

Table E–1. Daily Heat Loss—Average Outside Temperature 12.4° F

Daytime loss	(10 hours)	188,300
Nighttime loss	(14 hours)	147,800
Internal gain		(60,000)
Total Net Loss (Btu)		276,100

Table E–2. Monthly Heating Energy Use During the Minneapolis Heating Season*

		Conventional House	ESSPS House		
	Av. Temp. (F)	Net Heat Loss	Net Heat Loss	Exchanger Energy**	Total ESSPS Energy
Sept.	60.4	3.6	—	—	—
Oct.	48.9	10.0	1.9	0.5	2.4
Nov.	31.2	19.2	4.9	0.5	5.4
Dec.	18.1	27.1	7.5	0.5	8.0
Jan.	12.4	30.8	8.6	0.5	9.1
Feb.	15.7	25.9	7.2	0.4	7.6
Mar.	27.4	22.5	5.8	0.5	6.3
Apr.	44.3	12.2	2.8	0.5	3.3
May	57.3	5.3	1.7	—***	1.7
Totals		156.6	40.4	3.4	43.4

* In MBtu.

** Computed at local rate of 5.22¢ kwh.

*** Exchanger does not operate.

Table E–3. Clear-day Passive Gain and Average Daily Net Heat Loss in the Minneapolis Area ESSPS House*

	Clear-day Gain**	Average Daily Net Heat Loss	Clear-day Gain as % of Average Daily Net Loss
Oct.	527,600	60,200	876
Nov.	503,400	164,900	305
Dec.	435,600	242,400	180
Jan.	484,400	276,100	175
Feb.	548,900	256,500	214
Mar.	537,700	187,300	287
Apr.	446,300	87,400	511
May	347,900	53,800	647

* In Btu.
**Transmission for glass and skylights: 0.81.

Table E-2 gives the building's monthly and annual demand for heating energy.

Table E–4. Average Daily Passive Gain—Minneapolis Area ESSPS House

	Average Daily Passive Gain*	Average Daily Passive Gain as % of Average Daily Net Heat Loss
Oct.	337,400	560
Nov.	257,300	156
Dec.	225,000	93
Jan.	269,600	98
Feb.	334,800	131
Mar.	354,800	189
Apr.	341,500	390
May	344,800	640

Table E–5. Monthly Passive Solar Deficits—Minneapolis Area ESSPS House*

	Number of Deficit Days	Average Size of Deficit**	Monthly Deficit
Oct.	1	15,900	15,900
Nov.	11	140,900	1,549,900
Dec.	19	124,100	2,357,900
Jan.	18	127,200	2,289,600
Feb.	10	106,500	1,065,000
Mar.	7	80,700	564,900
Apr.	5	34,800	174,000
May	3	33,400	100,200

* In Btu.
** Average daily heating requirement less average daily gain during deficit days.

With 375 square feet of window area, and 40 square feet of nearly horizontal skylights, clear-day gain substantially exceeds average daily heat loss throughout the heating season.

When average daily gain during the month is compared with average daily net heat loss, the figures for gain and loss are much closer, and in December and January average net gain drops slightly below average net heat loss.

This creates a relatively large number of deficit days during the winter months, but total monthly deficits remain comparatively small.

The storage system has the capacity to keep the house warm overnight when temperatures drop as low as −45° F. For average nighttime temperatures, heat remaining at 7 A.M., when the storage system has been brought up to full capacity the night before, is shown in table E-6.

The pattern of weather variation in the Minneapolis region is broadly similar to that of the Philadelphia area. Energy stored in the passive system will be used at least four and one-half times each month to offset daily deficits, providing 16,000 to 148,000 Btu on each occasion.

Nearly 150 MBtu of heating energy is saved, a much higher total than the savings for the Philadelphia area house. Added construction cost for the Minneapolis area house, compared with the Philadelphia area house, is $1,325, of which $975 is for additional passive storage capacity, and $350 for extra insulation. The addition to the sale price of the house is $1,600, most of which qualifies for the solar energy tax credit. The Minneapolis area house obtains 65 percent greater savings in winter heating costs. When the lower electrical energy costs of the Minneapolis area are taken into account, overall savings in energy cost are 37 percent greater than those obtained in the Philadelphia area, while modifications in construction result in only a small addition to the cost of the house. Overall, the ESSPS house, built in a northern climate, becomes substantially more cost-effective than it is in a climate with moderate winter temperatures.

Table E–6. Nighttime Heat Loss and Passive Solar Storage Capacity*

	Av. Nighttime Outdoor Temp.**	Av. Nighttime Heat Loss	Av. Nighttime Net Heat Loss***	Av. Stored Heat at 7 A.M.****
Oct.	47.1	49,900	19,900	193,100
Nov.	29.4	86,800	56,800	156,200
Dec.	16.3	114,100	84,100	128,900
Jan.	10.6	126,000	96,000	117,000
Feb.	13.9	119,100	89,100	123,900
Mar.	25.6	94,700	64,700	148,300
Apr.	42.5	59,500	29,500	183,500
May	55.5	32,300	2,300	210,700

* In Btu.
** Minneapolis area.
*** 30,000 Btu nighttime internal gain.
**** Storage capacity: 213,000 Btu.

Table E–7. Monthly Passive Deficit and Estimated Offset From Storage in the Minneapolis Area ESSPS House*

	Monthly Passive Deficit	Energy Supplied by Storage System During Deficit Days	Remaining Energy Deficit
Oct.	16,000	16,000	—
Nov.	1,550,000	703,000	847,000
Dec.	2,358,000	580,000	1,778,000
Jan.	2,290,000	526,000	1,764,000
Feb.	1,065,000	558,000	507,000
Mar.	567,000	567,000	—
Apr.	174,000	174,000	—
May	100,000	100,000	—
Totals	8,120,000	3,224,000	4,896,000

* In Btu.

Table E–8. Winter Space-heating Energy Demand for the Minneapolis Area ESSPS House*

Conventional House	ESSPS House			
Net Heat Loss	Net Heat Loss	Heat Exchanger Energy Use	Passive Input	Total** Heating Energy
156.6	40.4	3.4	35.5	8.3

* In MBtu.
** Used by backup system and exchanger.

INDEX